Advances and Applications of Rheology

Advances and Applications of Rheology

Edited by **Eldra Lipton**

New York

Published by NY Research Press,
23 West, 55th Street, Suite 816,
New York, NY 10019, USA
www.nyresearchpress.com

Advances and Applications of Rheology
Edited by Eldra Lipton

International Standard Book Number: 978-1-63238-027-2 (Hardback)

Printed in the United States of America.

Contents

Preface VII

Chapter 1 **A Practical Review of Microrheological Techniques** 1
Bradley W. Mansel, Stephen Keen, Philipus J. Patty, Yacine Hemar and Martin A.K. Williams

Chapter 2 **Heliogeophysical Aspects of Rheology: New Technologies and Horizons of Preventive Medicine** 22
Trofimov Alexander and Sevostyanova Evgeniya

Chapter 3. **Rheological Characterisation of Diglycidylether of Bisphenol-A (DGEBA) and Polyurethane (PU) Based Isotropic Conductive Adhesives** 40
R. Durairaj, Lam Wai Man, Kau Chee Leong, Liew Jian Ping, N. N. Ekere and Lim Seow Pheng

Chapter 4 **Performance of Fresh Portland Cement Pastes - Determination of Some Specific Rheological Parameters** 56
R. Talero, C. Pedrajas and V. Rahhal

Chapter 5 **Unsteady Axial Viscoelastic Pipe Flows of an Oldroyd B Fluid** 79
A. Abu-El Hassan and E. M. El-Maghawry

Chapter 6 **Measurement and Prediction of Fluid Viscosities at High Shear Rates** 95
Jeshwanth K. Rameshwaram and Tien T. Dao

Permissions

List of Contributors

Preface

The world is advancing at a fast pace like never before. Therefore, the need is to keep up with the latest developments. This book was an idea that came to fruition when the specialists in the area realized the need to coordinate together and document essential themes in the subject. That's when I was requested to be the editor. Editing this book has been an honour as it brings together diverse authors researching on different streams of the field. The book collates essential materials contributed by veterans in the area which can be utilized by students and researchers alike.

This book collects current research in the field of rheology. It covers various topics related to this field. Polymer gels, liquid crystals, food rheology, etc. are some of the subjects discussed within the book. This book will be helpful for researchers, scientists, academicians and industrial experts.

Each chapter is a sole-standing publication that reflects each author´s interpretation. Thus, the book displays a multi-facetted picture of our current understanding of application, resources and aspects of the field. I would like to thank the contributors of this book and my family for their endless support.

Editor

Chapter 1

A Practical Review of Microrheological Techniques

Bradley W. Mansel, Stephen Keen, Philipus J. Patty,
Yacine Hemar and Martin A.K. Williams

Additional information is available at the end of the chapter

1. Introduction

Microrheology is a method for the study of the viscoelastic properties of materials [1, 2]. It has many potential benefits including requiring only microlitres of sample and applying only microscopic strains, making it ideal for costly, rare or fragile samples. Ever since the earliest papers began emerging in the biophysical arena some ten to fifteen years ago [3,4], to more current publications [5-8] fascinating insights into the material properties of the cell and its constituent biopolymers have been revealed by microrheological studies. It can extract information about the underlying heterogeneities in soft materials of interest, and can measure viscoelastic properties to high frequencies compared to traditional rheological measurements [9]. This paper reviews the limits of speed and accuracy achievable with current advances in instrumentation, such as state-of-the-art correlators and cameras, by directly comparing different methodologies and equipment.

2. Basic principles

2.1. Extracting traditional rheological parameters

To use microrheology to obtain the traditional storage and loss moduli, (G′, G′′), of complex soft materials of interest, the mean square displacement (MSD) of microscopic tracer particles must be measured, defined in three dimensions as:

$$\left\langle \Delta r^2(\tau) \right\rangle = \left\langle \left[x(t+\tau)-x(t)\right]^2 + \left[y(t+\tau)-y(t)\right]^2 + \left[z(t+\tau)-z(t)\right]^2 \right\rangle \quad (1)$$

where, τ is the lag time, t is the time and x, y and z represent position data [10]. There are a number of experimental techniques to measure the MSD, each with its own advantages and disadvantages that will be described in due course.

If a material is purely viscous, the MSD of an ensemble of thermally-driven tracer particles will increase linearly with time, yielding a logarithmic plot having a slope of one. In contrast, tracers embedded in a purely elastic material will show no increase in the MSD with time and the particle's location will simply fluctuate around some equilibrium position. While these two limiting cases are intuitive many materials of interest, particularly in the biophysical arena, are viscoelastic, both storing and dissipating energy as they are deformed. This is signaled by a slope between the extreme cases of zero and one on a logarithmic plot of MSD versus time. Additionally materials often display differing viscoelastic properties on different time-scales so that the slope of such a plot can change throughout the experimentally observed range. Indeed, the range of lag times over which the MSD is measured is equivalent to probing the viscoelastic properties as a function of frequency. Whilst the basic idea of using the dynamic behavior of such internal colloidal probes as an indication of the viscoelasticity of the surrounding medium has a long history, it took the relatively recent availability of robust numerical methods to transform the raw MSD versus time data into traditional viscoelastic spectra to drive the field forwards [10].

Tracer particles embedded in a **purely viscous medium** have an MSD defined by:

$$\left\langle \Delta r^2(\tau) \right\rangle = 2dD\tau \tag{2}$$

where τ is the lag time, d is the dimensionality and D is the diffusion coefficient, which is defined by the ratio of thermal energy to the friction coefficient, as embodied by the famous Einstein-equation:

$$D = \frac{k_B T}{f} \tag{3}$$

where, T, is the temperature and f is the friction coefficient. For added spherical tracers in low Reynolds number fluids, f can be calculated by the Stokes drag equation for a sphere:

$$f = 6\pi\eta R \tag{4}$$

where η is the viscosity of the surrounding material and R, the radius of the tracer.

Tracer particles embedded in a **viscoelastic medium** do not have such a simple relation between the MSD and diffusion coefficient. However, a Generalized Stokes-Einstein Relation (GSER) can be used, that accommodates the viscoelasticity of a complex fluid as a frequency dependent viscosity, yielding [1, 10, 11]:

$$\tilde{G}(s) = \frac{k_B T}{\pi a s \left\langle \tilde{r}^2(s) \right\rangle} \tag{5}$$

where $\langle \tilde{r}^2(s) \rangle$ is the Laplace transform of the MSD and, $\tilde{G}(s)$ is the viscoelastic spectrum as a function of Laplace frequency, s [1]. This relationship provides a method to quantify the rheological properties of a viscoelastic medium and calculate the storage and loss modulus from the MSD measurement. Many methods are available to implement this scheme, although the numerical method of Mason and Weitz is possibly the most popular method, due to its simplicity and ability to handle noise [10]. Briefly, the MSD plot is fitted to a local power law and the logarithmic differential is then calculated:

$$\alpha(\tau) = \frac{d \ln \left\langle \Delta r^2(\tau) \right\rangle}{d \ln(\tau)} \tag{6}$$

which is used with, Γ, the gamma function in an algebraic form of the GSER:

$$\left| G^* \right| \approx \frac{k_B T}{\pi a \left\langle \Delta r^2 (\tau = 1/\omega) \right\rangle \Gamma\left[1 + \alpha(\tau = 1/\omega)\right]} \tag{7}$$

Finally, defining $\delta(\omega)$ as:

$$\delta(\omega) = \frac{\pi}{2} \frac{d \ln \left| G^*(\omega) \right|}{d \ln \omega} \tag{8}$$

then the storage and loss moduli with respect to frequency can be obtained:

$$G'(\omega) = \left| G^*(\omega) \right| \cos(\delta(\omega)) \tag{9}$$

$$G''(\omega) = \left| G^*(\omega) \right| \sin(\delta(\omega)) \tag{10}$$

Thus, with the framework of microrheology clear and modern methods in place to obtain traditional viscoelastic spectra from the movement of internalized tracer particles, the discussion switches to reviewing experimental methods for the extraction of their mean squared displacement.

2.2. Measuring the MSD

In order to facilitate the review of the available techniques four different modern techniques have been used to measure the positions of micron sized particles embedded in soft materials, namely: Dynamic Light Scattering (DLS), Diffusing Wave Spectroscopy (DWS), Multiple Particle Tracking (MPT), and probe laser tracking with a Quadrant Photo Diode (QPD) and the use of Optical Traps (OT).

2.3. Light scattering techniques

Dynamic light scattering (DLS) techniques for microrheology use a coherent monochromatic light source and detection optics to measure the intensity fluctuations in light scattered from tracer particles of a known size, which are embedded in a material of unknown viscoelastic properties. Light passing through the sample produces a speckle pattern that fluctuates as the scattering probe moves. Thus, by measuring the intensity fluctuations of the dynamic speckle, at a single spatial position, information about the diffusion of particles in the sample can be gathered [12]. A correlation function is defined by:

$$g^{(2)}(\tau) = \frac{\langle I(t)I(t+\tau)\rangle}{\langle I(t)\rangle^2} \tag{11}$$

With τ, the lag time, t the time and the angular bracket denoting a time average. For ergodic samples the auto-correlation function can be simply converted to the so-called field auto-correlation function, $g^{(1)}$, using the Siegert relation:

$$g^{(2)}(\tau) = 1 + \beta\left|g^{(1)}(\tau)\right|^2 \tag{12}$$

The coherence factor, β, in this relationship, is related to the experimental setup, and for a properly aligned system should be close to unity. DLS uses a sample containing a low number of probe scatterers to ensure that each photon exiting the sample has been scattered only a single time. Using recently developed techniques such as multiple scattering suppression [13] one can still extract some information if multiple scattering cannot be avoided, but these are not commonly used as sample optimization can often provide a simpler solution. Central to DLS experiments is the scattering vector defined by:

$$q = \frac{4n\pi}{\lambda}\sin\left(\frac{\theta}{2}\right) \tag{13}$$

where λ represents the wavelength of the incident laser light, n, the refractive index of the medium surrounding the scatterer and θ the angle the incident beam makes with the detector. Ultimately, for traditional DLS experiments, the q vector must be known to extract information about the displacements made by the particles. For more information see Dasgupta [14] or Pecora [12].

Practically, light emitted from a continuous wave, vertically-polarized laser is directed through the sample held in a goniometer. Using a polarized laser combined with a crossed polarizer on the detection optics helps to reduce the chance of light that has not been scattered entering the detection optics, which helps improve the signal. As well as providing angular control the goniometer typically has a bath surrounding the sample that is filled with a fluid of a similar refractive index to the cuvette in which the sample is housed, to help eliminate light reflections from the surface. In the case of the DLS setup used in our studies, detection optics in the form of a gradient index (GRIN) lens directs photons scattered at a particular angle into a single-mode optical fiber that incorporates a beam splitter. The two beams thus produced are taken to two different photo multiplier tubes (PMTs) that produce electronic signals. These are interrogated by a correlator interfaced to a computer that converts fluctuations in the scattered light falling onto the PMTs into a correlation function. When two photomultiplier tubes are used the *cross*-correlation function can be formed, as opposed to an *auto*-correlation function that can be measured with a single PMT. Cross correlation help circumvent dead time in the electronics as well as helping eliminate after-pulsing effects. A schematic of a typical experimental setup is shown in figure 1(a).

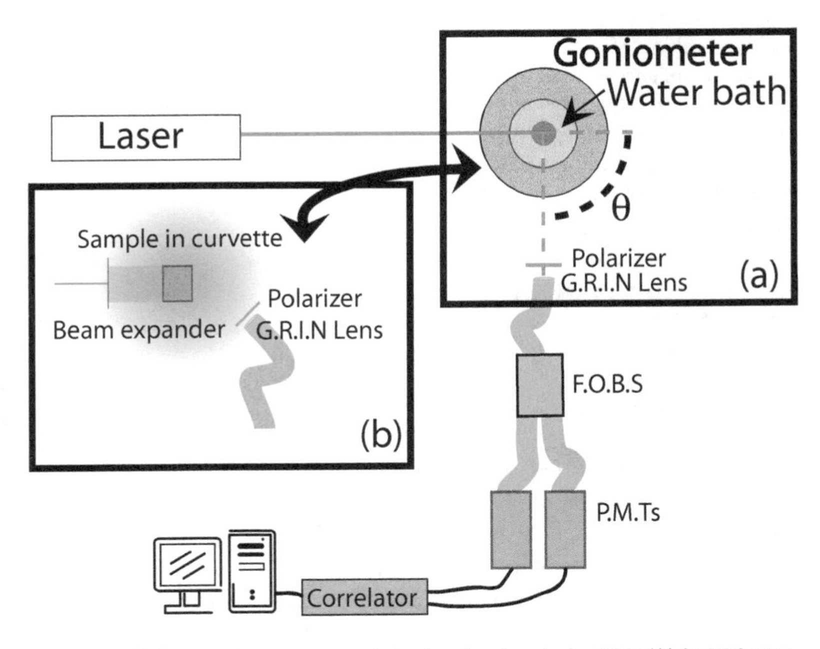

Figure 1. Schematic of light scattering apparatus used, showing a) goniometer for DLS and b) the DWS setup.

In DLS, where single scattering events dominate, the decay of the field correlation function, $g^{(1)}$, is related to the diffusion of the particles in the sample by:

$$g^{(1)}(\tau,q) = \exp(-Dq^2\tau) \tag{14}$$

where τ represents the lag time, q, the scattering vector and D the diffusion coefficient which, by equation (2) can be written as [12]:

$$g^{(1)}(\tau,q) = \exp\left(\frac{-q^2\left\langle \Delta r^2(\tau)\right\rangle}{6}\right) \tag{15}$$

By inverting this equation one obtains the MSD versus lag time directly from the field correlation function.

Diffusing Wave Spectroscopy: At high frequencies DLS is limited by the sensitivity of the correlator. This limitation can be overcome by adding many scatterers to the sample. The light now diffuses through the sample taking a random walk with mean-free path, l [15]. The diffusion of light through the sample means that even if each individual scatterer was only to move a very small amount, the overall path that the light travels is changed very dramatically, resulting in a much higher sensitivity than DLS. However, when making measurements in materials with a very large number of scatterers a statistical approach must be used to derive the form of the correlation function. To ensure the accuracy of the statistical approach the number of scatterers must be large enough so that the photon paths can be themselves described by a random walk. Light scattering in this high scattering limit is known as Diffusing Wave Spectroscopy (DWS) [15].

The equipment used for DWS is very similar to that used for DLS. The main difference is that no goniometer is required as, provided that all the photons studied have traversed the cell, there is no angular dependence of the intensity of scattered light. Additionally the incident beam is first expanded to distribute the intensity of the light across the width of the sample cuvette, (in the case described here to around 8 millimetres). Otherwise, as in DLS, a continuous wave, vertically polarized, laser is used as the light source; the scattered light is coupled to a single mode optical fibre using a GRIN lens; split by a single mode fiber-optic beam-splitter (FOBS) and sent to two PMTs. The correlation function is then calculated using a cross-correlation method in software on a standard personal computer. A schematic of the experimental setup used here can be seen in figure 1 (b).

The measurement of the length a photon must travel before its direction is completely randomized, l^*, is fundamental to DWS. Firstly, comparing it to the pathlength of the cell reveals if the number of scatterers present in a sample is large enough to validate the diffusive criterion, and secondly, it is needed in order to extract the MSD from the correlation function. l^*, being at least four times smaller than the thickness of the sample ensures that the light is strongly scattered, producing statistically viable results. To calculate the MSD in transmission geometry, with uniform illumination over the face of the sample, an inversion is performed on the following equation [15]:

$$g^{(1)}(\tau)=\frac{\left(\frac{L/l^{*}+4/3}{z_0/l^{*}+2/3}\right)\sinh\left(\frac{z_0}{l^{*}}\sqrt{k_0^2\langle r^2\rangle}\right)+\frac{2}{3}\sqrt{k_0^2\langle r^2\rangle}\cosh\left(\frac{z_0}{l^{*}}\sqrt{k_0^2\langle r^2\rangle}\right)}{\left(1+\frac{8t}{3\tau}\right)\sinh\left(\frac{L}{l^{*}}\sqrt{k_0^2\langle r^2\rangle}\right)+\frac{4}{3}\sqrt{k_0^2\langle r^2\rangle}\cosh\left(\frac{L}{l^{*}}\sqrt{k_0^2\langle r^2\rangle}\right)} \tag{16}$$

Here L represents the thickness of the sample and l^{*} the transport mean free path of the medium. It is assumed that the source of diffusing intensity is a distance, z_0, inside the sample, which is routinely assumed to be equal to l^{*}. k_0 is the wave vector of the incident light, equal to $2\pi\lambda$. For a detailed description on the theory of DWS and the mathematics behind equation (16) see chapter 16 of *Dynamic light scattering: The Method and Some Applications* edited by Wyn Brown, which covers this extensively [15].

Summary: Generally light scattering techniques have the advantage that they have a low setup cost, are well known and produce reliable results. DLS, one of the more common light scattering techniques, cannot measure to the high frequencies of DWS and also has a slightly higher setup cost as a goniometer is required for angular control. However, if many measurements can be taken and averaged, DLS can produce very consistent results, and if the scattering angle is reduced it is possible to obtain particle dynamics out to tens of seconds. DWS is also a robust, well proven technique that can achieve higher frequency measurements than any other method, due to a small displacement of the bead causing an additive effect in each successive scatter through the sample. Traditional light scattering experiments do not however have the ability to extract any information about the homogeneity of the sample, although this can be accomplished to some extent using modified techniques such as multispeckle DWS, where the PMT is replaced by a camera [13, 16].

2.4. Real space tracking techniques

Multiple particle tracking typically consists of visually tracking tens to hundreds of probe particles embedded in the material to be studied [9]. Commonly, an epifluorescence microscope with a CCD or CMOS camera is used to record a series of images of fluorescent tracer particles as they undertake random walks due to Brownian motion. Fluorescence microscopy has many advantages over simple bright field microscopy; it produces images with the particles represented as bright spots on a dark background, facilitating the use of many different tracking algorithms, and allows the position of particles smaller than the wavelength of light to be obtained. Image series taken from the chosen microscopy technique are subsequently processed using tracking software, turning the images into a time-course of x-y coordinate data for each particle. From this data the MSD can be calculated and hence the rheological information extracted. MPT is mainly limited by the temporal resolution of the camera (typically 45 Hz), meaning that lower frequency rheological information is accessible when compared with other light-scattering based microrheology techniques. It has advantages however of measuring information about the spatial homogeneity of the sample, and is one of the few techniques capable of studying the viscoelastic properties of living samples where, for example, naturally occurring particles (liposomes and organelles) might be tracked.

Information about spatial homogeneity:
A plot of the probability of displacements of a certain value observed at each time lag gives an indication of homogeneity. If the sample is homogenous then one would expect this to result in a Gaussian function centred on the origin at each time lag, with the variances of the distributions being the MSDs. The plotting of the frequency with which a measured displacement falls in a particular displacement-range is known as a Van Hove plot [17], and is shown for particles diffusing in water in figure 2.

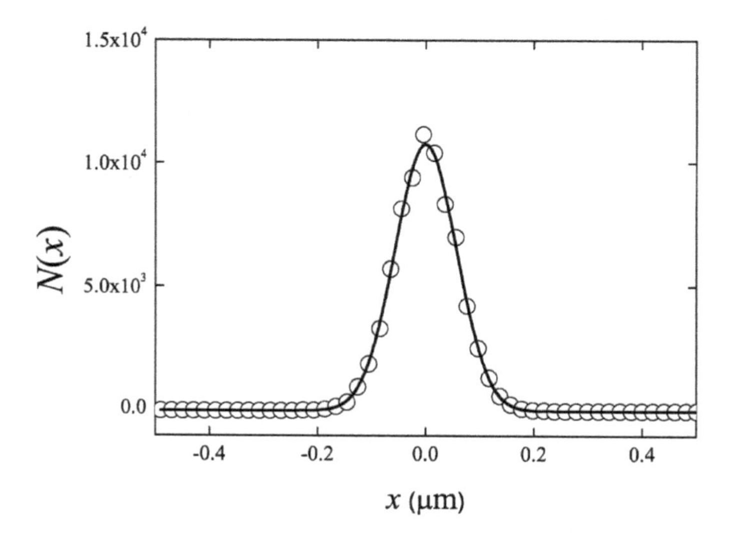

Figure 2. A Van Hove plot measured for 505nm polystyrene fluorescence particles diffusing in a glycerol water mixture, a homogenous medium, as can be seen by a good agreement to a Gaussian fit (solid line). Data obtained using a CMOS camera at 45Hz.

If significant heterogeneities exist then the Van Hove function will be non-Gaussian, indicating that the differences in the distances travelled by different beads in the same time does not simply represent the sampling of a stochastic process, but that differences in the local viscoelastic properties exist. The Van Hove plots of such heterogeneous systems can be quantified by a so-called non-Gaussian parameter that reports how much the ratio of second to fourth moments of the distribution differs from the Gaussian expectation [18, 19]. Information about the underlying structure of the sample can also be extracted by observing the behavior of the MSD when probe particles of different sizes are used [20].

The so-called one-point microrheology (OPM) described thus far simply extracts the displacements of each probe particle by comparing their co-ordinates in time-stamped frames recorded by the camera using a tracking algorithm. This is the simplest form of analysis and is often sufficient. However, the results can be highly dependent on the nature of interactions existing between the tracer particles and the medium, and effects of any specific binding, or depletion interactions can produce spurious measurements of the viscoelastic properties of the medium [21]. That is, OPM can be thought of as a superposition of the bulk rheology and the rheology of the material at the particle boundary [22]. With video-microscopy, where multiple probe-particles are tracked simultaneously, a method to overcome these difficulties has been developed, known as two-point microrheology (TPM). TPM only differs in the way the data is analysed, in that, rather than just looking at one particle TPM measures the cross-correlation of the movement of pairs of particles [22]. In some cases TPM has been shown to measure viscoelastic properties in better agreement with those measured using a bulk rheometer, due to the elimination of dependence on particle size, particle shape, and coupling between the particle and the medium [23]. TPM is a fairly intuitive technique if the two limiting cases are considered; the probe particles in an elastic solid will exhibit completely correlated motion throughout the sample, while in a simple fluid they would exhibit very little correlated motion. In between these extremes the viscoelasticity can be quantified by knowledge of the distance between particles, the thermal energy and the cross-correlation function [24]. While it does have potential advantages, two-point microrheology is very susceptible to any drift or mechanical vibration; which appears as completely correlated motion [24]. If the material of interest is homogenous, incompressible, isotropic on length-scales significantly smaller than the probe particle, and connected to the tracers by uniform no-slip boundary conditions over the whole surface, then the one- and two- point MSDs should be equal [24].

To perform two-point microrheology first the ensemble average tensor product is calculated:

$$D_{\alpha\beta}(r,\tau)=\left\langle \Delta r_{\alpha}^{i}\left(t,\tau\right)\Delta r_{\beta}^{i}\left(t,\tau\right)\delta\left(r-R^{ij}\left(t\right)\right)\right\rangle_{i\neq j,t} \tag{17}$$

where i and j label different particles, and label different coordinates, and R^{ij} is the distance between particle i and j. The distinct MSD can be defined by rescaling the two-point correlation tensor by a geometric factor [22-24]:

$$\left\langle \Delta r^{2}\left(\tau\right)\right\rangle_{D}=\frac{2r}{a}D_{rr}\left(r,\tau\right) \tag{18}$$

where a is the diameter of the probe particles. Further information on the mathematics behind the method can be obtained from Crocker (2007) [24] and Levine (2002) [23].

Tracking software: A plethora of different programs and algorithms exist to track objects in successive images. Both commercial and freeware programs exist. Commercial software such as Image Pro Plus, can track images straight out of the box with little fuss, although it is reasonably costly. One can also write their own program to cater to their own needs, and kindly many research groups have made free software available that generally works as well as many commercial packages. There are four main tracking algorithms, namely: centre of mass, correlation, Gaussian fit and polynomial fit with Gaussian weight. Ready to use programs are available on the following web pages:

http://www.physics.emory.edu/~weeks/idl/ This web page is a great resource with links to many different programs written in many different programming languages.

http://physics.georgetown.edu/matlab/ This code uses the centroid algorithm for sub-pixel tracking, it is the code used for the majority of the particle tracking in this work. Some knowledge of programming in MATLAB is needed to implement the code.

http://www.people.umass.edu/Kilfoil/downloads.html This resource has code available for calculating the MSD, two-point microrheology, and many other useful programs implemented in MATLAB.

http://www.mosaic.ethz.ch/Downloads/ParticleTracker This page has links to a 2D and 3D particle tracking algorithm, as published in [25]. The code is implemented using ImageJ a popular Java-based open source image processing and analysis program.

http://www.mathworks.de/matlabcentral/fileexchange/authors/26608 Polyparticle tracker uses a polynomial fit with Gaussian weight. This powerful tracking algorithm has a good graphical user interface and is easy to implement. Details of the algorithm can be viewed in the following publication [26].

Theoretically it can be seen that the selection of tracking algorithm could play a large role in multiple particle tracking experiments. In reality the differences in performance between different tracking algorithms can largely be overcome by optimizing the experimental setup. Indeed Cheezum (2001) [27] have shown that at a high signal-to-noise ratio the different tracking algorithms produce very similar bias and standard deviations. The lower limit where differences in the algorithms do become important is a signal-to-noise of around 4, which roughly corresponds to imaging single fluorescent molecules. The fluorescent microspheres imaged in multiple particle tracking experiments are many tens of times brighter than the background fluorescence, generally providing a high signal-to-noise. Additionally, modern cameras have photo-detector arrays consisting of many megapixels, resulting in a particle diameter in the order of tens of pixels, so that effects from noise on the edge of a particle often have little effect. Oscillation and drift in an experimental setup can however create large sources of error and often are the hardest errors to remove. For a more in depth description see references Rogers (2007) [26] and Cheezum (2001) [27]. Errors in particle tracking can be placed in 4 different categories: Random error, systematic error, dynamic error and sample drift. A thorough discussion is given in Crocker (2007) [24] and further precise methods with which to estimate the static and dynamic errors present in particle tracking are given by Savin (2005, 2007) [28, 29]. Practically multiple experimental techniques are often used and the comparison of results quickly reveals if significant errors in the MPT are present.

Optimizing experimental set-up for microscope based experiments: The camera used for MPT is the central apparatus limiting the temporal and spatial resolution. Current CMOS technology allows the fastest frame rate of any off-the-shelf camera designed for microscopy [30, 31]. The main problem with this technology is the sensitivity, although these issues are beginning to be addressed [32]. Cameras with a high sensitivity, large detector size, higher speed and small pixel size can obtain a larger amount of information from the sample. To supply the tracking algorithm with enough information to calculate the position of a probe particle to sub-pixel accuracy, the particle must be represented by a sufficient number of pixels. The size representation of fluorescent particles is dependent on the size of the particles, the intensity of the excitation fluorescent lamp, the size of each pixel on the sensor, and the magnification of the objective lens used. The strength of the fluorescent lamp that can be used is ultimately limited by the speed at which it photo-bleaches the fluorophore. One possible method to overcome the photo-bleaching difficulties is to use quantum dots.

Magnification: For high signal-to-noise applications, it is advantageous to have the highest possible magnification, resulting in the particles being represented by the maximum number of pixels, and so enhancing the accuracy of the tracking algorithm subsequently applied [33]. However, as the magnification is increased the illumination of each pixel decreases as the square of the magnification. This results in a decrease in the signal-to-noise proportional to the magnification, if the illumination is not increased [33] and therefore for low signal-to-noise applications, the highest possible magnification will not always result in the best image sequence for tracking. As a result care must be taken in selecting the correct magnification objective lens. A simple method to check that the selected objective is of the correct magnification before recording an image sequence is to record a single image, then using an image analysis program such as ImageJ (http://rsbweb.nih.gov/ij/) to find the brightness of an individual pixel on a particle. This can then be compared to the background brightness of the image. By comparing the two intensity values one can roughly estimate the signal-to-noise. If the signal to noise is too low (< ~10) then a lower power objective lens can be chosen. This basic method will suffice to quickly give an indication of what objective is appropriate for the sample. A lower magnification objective will also result in a larger field of view in the sample, thus, the positions of more individual particles can be measured, and better statistics of the ensemble averaged MSD will result.

Numerical aperture: A high numerical aperture (NA) objective lens creates a higher resolution image than the equivalent lower NA objective lens. This would suggest that a high NA objective lens would create a superior image for tracking, although, a high NA objective also results in a small point-spread function, meaning a smaller image. In reality these two competing effects relating to the NA lens used usually cancel out. A simple calculation shows that if the signal-to-noise is high (around 30) then there is no effect of the NA used. No relation between the NA and accuracy of tracking was found in an experiment performed using different tracking algorithms and comparing data for a 0.6 NA and 1.3 NA lens [33].

Allan variance: If no drift is present in an experimental setup, a very unlikely situation, then the longer the experiment is run the better the accuracy of the measurement. However, if drift is present, as in nearly every experimental setup, running the experiment for the longest duration will not result in the highest accuracy measurement, it will actually result in a worse measurement than if the measurement was taken for a shorter duration. One can check the optimum length of time for which to record an experiment by using the Allan Variance. Defined as:

$$\sigma_x^2(\tau) = \frac{1}{2}\left\langle \left(x_{i+1} - x_i\right)^2 \right\rangle_\tau \tag{19}$$

where τ is the time lag, x_i is the mean over the time interval defined as $(\tau) = f_{acq}m$, where m is the number of elements in that interval acquired at f_{acq}, and the angle brackets denote arithmetic mean [34]. Most commonly the Allan variance is used in optical tweezers experiments, and the tracking performed using a Quadrant Photodiode (QPD) discussed in the following section, although with modern CMOS cameras approaching the kilohertz regime one can optimize particle-tracking experiments in this way. The Allan variance can be seen in figure 3 to decrease as the number of measurements (number of lag times evaluated) increases, until such long lag times are used that drift becomes a significant effect on the measurement.

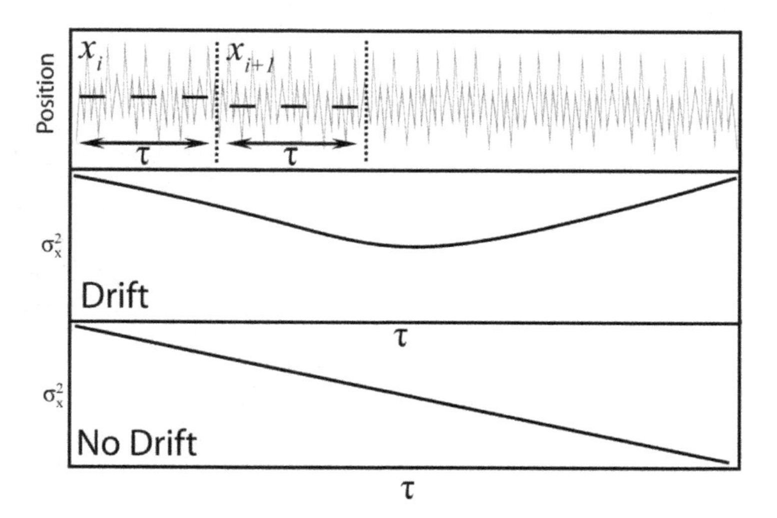

Figure 3. Schematic showing the relation between a particles mean position and the Allan Variance. It can be seen that in a perfect experiment, with no drift, the Allan variance decreases for the duration of the experiment, but where drift is present the Allan variance has a minimum corresponding to when the effects of sampling statistics and drift are balancing out.

2.5. QPD measurements using optical traps

The movement of individual probe particles can also be tracked using a probe laser and a quadrant photodiode (QPD) (a photodiode that is divided into four quadrants). A probe laser is used to scatter light from the selected particle and this produces an interference pattern that is arranged to fall on the QPD. Two output voltages are produced from the difference- signals generated by light falling on different quadrants and therefore any movement of the interference pattern on the QPD is detected by a change in output voltages. Thus, if the probe particle is located between the laser and QPD, any motion of the particle will be detected. Once calibrated these recorded voltages correspond directly to a measurement of the x and y co-ordinates of the probe particle - so that ultimately the output is equivalent to that which would be obtained by a video-microscopy tracking experiment. While the calibration requires an extra step in the measurements, a QPD has the advantage that measurements are not limited by a camera frame rate and for commercial QPDs can be taken on the order of tens of microseconds, subsequently giving access to rheological information up to the 100 kHz regime, albeit one probe particle at a time.

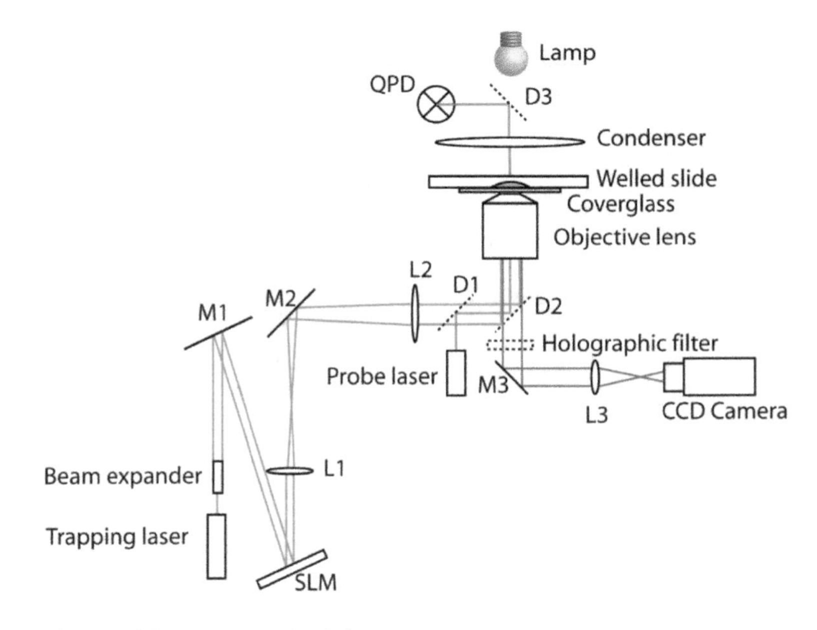

Figure 4. Schematic of the microscope and optical tweezers apparatus.

There is, however, an additional complication in making such measurements. Clearly a fixed QPD is limited in the maximum particle displacement it can measure and as probe particles are diffusing in 3 dimensions it is essential to provide a mechanism that ensures the particle being tracked stays within the range of the QPD. This can be carried out with an optical tweezers arrangement that uses a tightly focused higher-power laser to hold and manipulate micron-sized particles [35-37]. Figure 4 shows a typical holographic optical tweezer setup (HOT) that employs a spatial light modulator (SLM), which provides the ability to make multiple steerable traps and move objects in three dimensions using a single laser, in real time [38, 39].

Optical traps [40] formed by such an arrangement can be utilized to restrict larger-scale movements of probe particles so they stay in the detection region of the QPD / laser apparatus - essentially fencing them in, while leaving the smaller scale Brownian-motion unperturbed. At longer time lags the effect of the trap can be seen in the MSD plot, as a plateau indicating the effect of the trap, as shown in figure 5.

Calibration of the raw photodiode voltages in order to obtain actual bead displacements are routinely carried out by moving a probe particle a set distance across the QPD detection area. This can be carried out either by locating a particle that is stuck to the coverslip of the sample cell and translating the chamber a known amount using a piezo-electric stage; or by moving a particle using a pre-calibrated optical trap. An average piezoelectric stage currently available for microscopy is able to provide nanometer resolution to displacements up to 300 microns.

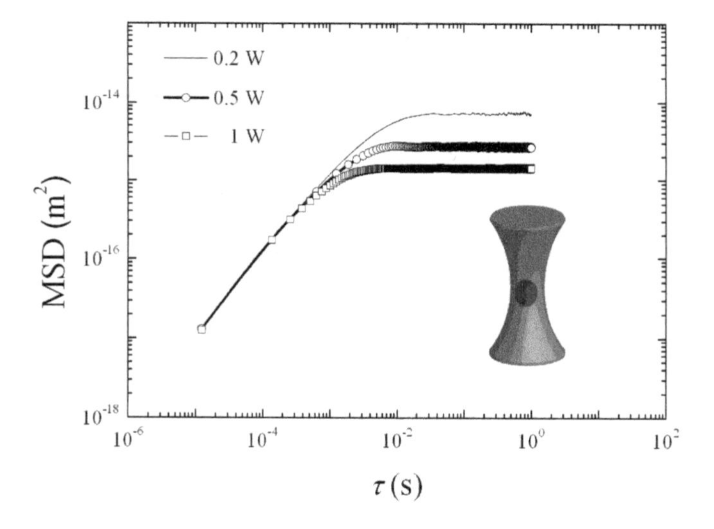

Figure 5. MSD plot of a particle undertaking Brownian motion within optical traps formed with three different laser intensities. The insert shows a particle optically trapped.

2.6. Standard experimental studies

Having described the setup and calibration of four microrheological techniques, results obtained from 3 different fluids are described and compared. Water, a glycerol-water mixture, and several polyethylene oxide (PEO) solutions were utilized to provide three different environments, namely; low viscosity, high viscosity and viscoelastic fluids, to test and compare the different methods. Such samples are standards that can be quickly used to ensure the proper functioning of the equipment and analysis before more complex biological systems are investigated.

Samples: *Water* has a lower viscosity than most biological materials of interest; and thus provides a good test of how the methodologies cope with fast particle dynamics. *Glycerol* is a homogenous, purely viscous fluid and was used in combination with water (results shown here for 62 wt%) to generate a highly viscous solution. Solutions were made by mixing glycerol (99.9% from Ajax Laboratory Chemicals) and MilliQ water, using a magnetic flea, for approximately 2 hours. *PEO*, an electrically-neutral water-soluble polymer available in a range of molecular weights was used to generate a viscoelastic polymer solution. PEO starts to exhibit viscoelasticity at concentrations higher than the overlap concentration (approximately 0.16 wt% for the 900 kDa PEO samples used in the following experiments). Solutions were made by adding dry PEO powder (Acros Organics) in MilliQ water, and then slowly mixing over approximately a 7 day period to help homogenize the solution. Solutions were prepared at 2.2 wt% and 4 wt%, around 14 and 25 times the overlap concentration, to ensure significant viscoelasticity [14]. The mesh size of PEO solutions at these concentrations have been calculated to be the order of a few nanometres. As there is little evidence of surface effects between the particles and these solutions, and the solution is then homogenous on the length scale smaller than the particle size, it was expected that the one-and two-point microrheology should produce very similar results, and as such the system forms an ideal test of those two methodologies.

Probe Particles: DWS measurements require a high bead concentration producing a turbid solution and ensuring strong multiple scattering. On the other hand, optical tweezers experiments and DLS measurements require that the concentration of particles in the solution is very low. For optical tweezers one must ensure that that there is only one particle present in the imaging plane, any more and there is a chance that a second additional particle might get sucked into the trap, and as discussed DLS requires that photons only be scattered a single time. For DLS and DWS polystyrene particles are chosen due to their low density and good scattering properties. Silica particles are used for optical tweezers due to the high refractive index of silica, which ensures a strong trapping force. DLS and DWS experiments were carried out for all samples with solid polystyrene probe particles (Polysciences) at concentrations of 0.01% and 1%, respectively. Solutions for optical tweezers and MPT experiments were made to concentrations of 10^{-6} % solid silica (Bangs Laboratories) and 10^{-3} % solid fluorescent polystyrene particles (Polysciences).

DLS experiments were performed using a set-up as shown in figure 1, specifically using a 35 milli-watt Helium Neon laser (Melles Griot) and a goniometer (Precision Devices) set nominally to measure a 90 degree scattering angle. Measurements were taken for approximately 40 minutes.

DWS experiments were performed using a set-up based on work originally published in [41] and as shown in figure 1. Initially experiments were conducted using a flex99 correlator from correlators.com and a 35 milli-watt Helium Neon laser (Melles Griot). In the quest for shorter lag times and higher accuracy a flex02 correlator (correlator.com) was purchased. Experiments were first run using water to obtain l^* of the standard solution, and then repeated on the sample solution containing the same phase volume of scatterers. DWS experiments were typically run for approximately 40 minutes to one hour.

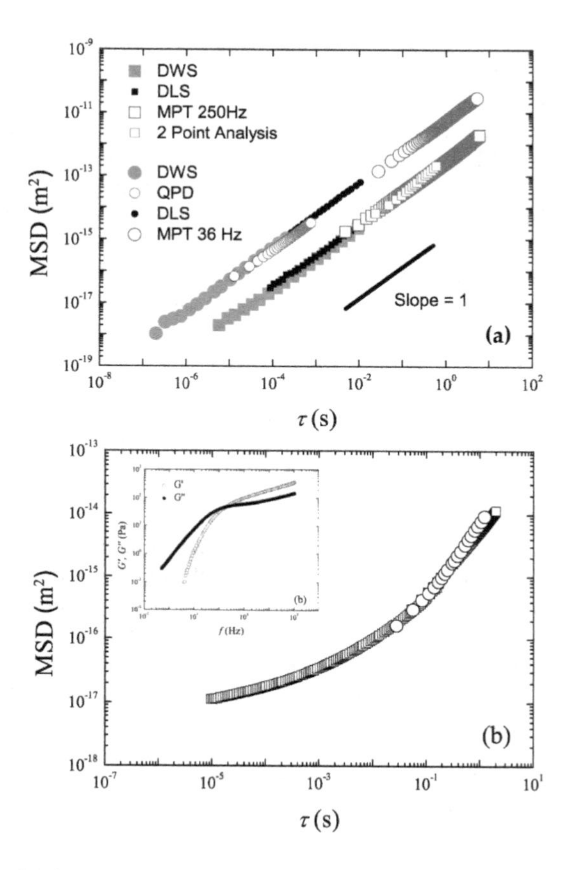

Figure 6. (a) Plot showing the agreement of measurements between multiple techniques in water (circles) and 62% glycerol water mixture (squares). (b) MSD plot for 4 wt% PEO showing an agreement between data obtained using DWS MPT and 2 point analysis (Inset: Extracted rheological properties).

MPT experiments were carried out with an inverted microscope (Nikon Eclipse TE2000-U) on an air damped table (Photon Control) equipped with a mercury fluorescent lamp (X-cite Series 120PC EXFO), and a 60x 1.2 NA (Nikon, Plan Apo VC 60x WI) water immersion objective lens was used for MPT experiments. A range of different cameras were trialled: Foculus FO124SC (CCD), prototype DSI-640-mt smartcam (high speed CMOS), Hamamatsu Orca Flash 2.8 (CMOS large detector size and pixel number). Image series were taken for approximately ten seconds; and x-y coordinate data extracted using a homebuilt program written using algorithms obtained from: http://physics.georgetown.edu/matlab/. In-house programs to calculate the MSD and Van Hove correlation function were used in combination with a program to extract the rheological information obtained from: http://www.physics.mcgill.ca/~kilfoil/downloads.html.

QPD experiments were also carried out. The microscope used for MPT was additionally utilized to tightly focus a 2 watt 1064 nm Nd:YAG laser (spectra physics) to produce optical traps. Particle displacements were recorded using a 2.5 mW probe laser (Thorlabs S1-FC-675) and a QPD (80 kHz) for approximately 10 seconds. Calibration was aided using piezoelectric multi-axis stage (PI P-517.3CD).

Figure 6(a) shows a log-log plot of the three dimensional mean-square displacement of probe-particles as a function of time; for 500 nm polystyrene particles and 1.86 micron silica beads (optical tweezers data, normalized to 500nm) in either water or a 62 wt% glycerol/water mixture. The mean-square displacement data shown shows an excellent agreement between different methods and also with the expected result of a slope of one (for diffusion in a viscous medium). Figure 6 (b) shows a similar log-log plot of the mean-square displacement versus time for 4 wt%, PEO solutions, together with a fit to a sum of power laws with exponents of ~0.4 and ~0.9, in good agreement with previous work. The inset shows the extracted frequency dependent viscoelastic properties that appear in good agreement with previously published work [14].

2.7. Comparison of with bulk rheometry

The efficiency of these microrheological methods can be assessed against conventional rheometry. Figure 7 reports the elastic modulus G' and the loss modulus G'' for a 30 wt% aqueous dextran solution. G' and G'' were obtained using DWS or by the use of a commercial rheometer (TA 2000 rheometer, fitted with a cone-and-plate geometry).

The experimental data were fitted using the Maxwell model:

$$G' = \frac{G_0 \omega^2 \tau^2}{1+\omega^2 \tau^2} \quad ;G' = \frac{G_0 \omega \ \tau}{1+\omega^2 \tau^2} \tag{20}$$

where G_0 is the plateau elastic modulus, τ is the relaxation time, and ω (ω=2 f, f the frequency) the angular frequency. The fit using the Maxwell model allows showing the continuation in the experimental data using the two methods. Further the combination of the two techniques allows the determination of the rheological behaviour over more than 7 decades in frequency. However, at low frequencies, some discrepencies between G' obtained by rheology and the Maxwell model can be observed. This is likely due to the geometry inertia affecting rheological measurements.

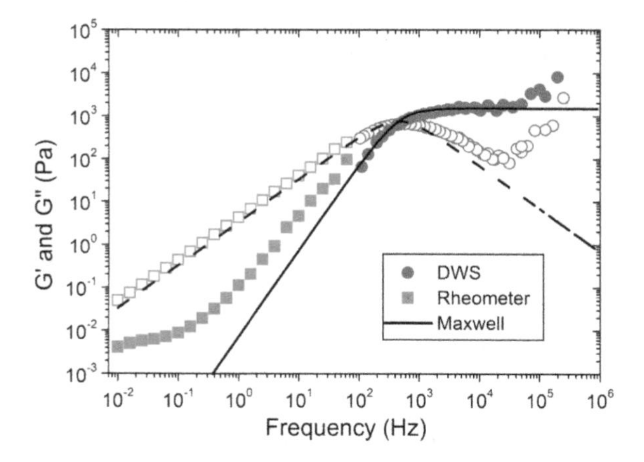

Figure 7. Elastic modulus G′ and loss modulus G″ as a function of frequency for a 30 wt% dextran (500 kDa) in water solution. Experimental data are obtained by conventional rheometry and DWS. Solid lines are a fit using a Maxwell model with one element.

3. Conclusion

The array of microrheology techniques described here provide the ability to measure the viscoelastic properties of a material over approximately nine orders of magnitude in time. The most sensitive technique, DWS, measured particle displacements to nanometre resolution, while MPT could measure the largest displacements, on the order of micrometres. Each technique can be used to measure the mechanical properties of both viscous and viscoelastic materials and has a promising future in experimental biophysics.

Author details

Bradley W. Mansel[1], Stephen Keen[1,2], Philipus J. Patty[1], Yacine Hemar[2,3] and Martin A.K. Williams[1,2,4]

1 Institute of Fundamental Sciences, Massey University, Palmerston North, New Zealand

2 MacDiarmid Institute for Advanced Materials and Nanotechnology, New Zealand

3 School of Chemical Sciences, University of Auckland, New Zealand

4 Riddet Institute, Palmerston North, New Zealand

References

[1] MacKintosh FC, Schmidt CF. Microrheology. Current Opinion in Colloid & Interface Science, 1999; 4 (4) 300-307.

[2] Gardel ML, Valentine MT, Weitz DA. Microscale diagnostic techniques. Springer, 2005.

[3] Yamada S, Wirtz D, Kuo SC. Mechanics of living cells measured by laser tracking microrheology. Biophysical journal 2000; 78 (4) 1736-1747.

[4] Tseng Y, Lee JSH, Kole TP, Jiang I, Wirtz D. Micro-organization and visco-elasticity of the interphase nucleus revealed by particle nanotracking. Journal of Cell Science. 2004; 117 (10):2159-2167. doi:10.1242/jcs.01073.

[5] Duits MHG, Li Y, Vanapalli SA, Mugele F. Mapping of spatiotemporal heterogeneous particle dynamics in living cells. Physical Review E 2009; 79 (5). doi:10.1103/PhysRevE.79.051910.

[6] Zhu X, Kundukad B, van der Maarel JRC. Viscoelasticity of entangled lambda-phage DNA solutions. Journal of Chemical Physics 2008; 129 (18). doi:185103 10.1063/1.3009249.

[7] Ji L, Loerke D, Gardel M, Danuser G. Probing intracellular force distributions by high-resolution live cell imaging and inverse dynamics. In: Wang YLDDE (ed) Cell Mechanics, vol 83. Methods in Cell Biology 2007, pp 199-+. doi:10.1016/s0091-679x(07)83009-3.

[8] Cicuta P, Donald AM. Microrheology: a review of the method and applications. Soft Matter 207; 3 (12):1449-1455. doi:10.1039/b706004c.

[9] Waigh TA. Microrheology of complex fluids. Reports on Progress in Physics 2005; 68 (3):685-742. doi:10.1088/0034-4885/68/3/r04.

[10] Mason TG. Estimating the viscoelastic moduli of complex fluids using the generalized Stokes-Einstein equation. Rheologica Acta 2000; 39 (4):371-378.

[11] Mason TG, Ganesan K, vanZanten JH, Wirtz D, Kuo SC. Particle tracking microrheology of complex fluids. Physical Review Letters 1997; 79 (17):3282-3285. doi:10.1103/PhysRevLett.79.3282.

[12] Pecora R. Dynamic light scattering: applications of photon correlation spectroscopy. Plenum Press, New York, 1985.

[13] Zakharov P, Bhat S, Schurtenberger P, Scheffold F. Multiple-scattering suppression in dynamic light scattering based on a digital camera detection scheme. Applied Optics 2006; 45 (8):1756-1764. doi:10.1364/ao.45.001756.

[14] Dasgupta BR. Microrheology and Dynamic Light Scattering Studies of Polymer Solutions. PhD Thesis; Harvard University, Cambridge, Massachusetts, 2004.

[15] Weitz DA, Pine DJ. Diffusing-wave spectroscopy. In: Brown W (ed) Dynamic Light Scattering: The method and some applications. Oxford University Press, Oxford, 1993; pp 652-720.

[16] Brunel L, Dihang H. Micro-rheology using multi speckle DWS with video camera. Application to film formation, drying and rheological stability. In: Co A, Leal LG, Colby RH, Giacomin AJ (eds) Xvth International Congress on Rheology - the Society of Rheology 80th Annual Meeting, Pts 1 and 2, vol 1027. Aip Conference Proceedings. pp 1099-1101, 2008.

[17] Valentine MT, Kaplan PD, Thota D, Crocker JC, Gisler T, Prud'homme RK, Beck M, Weitz DA. Investigating the microenvironments of inhomogeneous soft materials with multiple particle tracking. Physical Review E 2001; 64 (6). doi:061506 10.1103/PhysRevE.64.061506.

[18] Oppong FK, Rubatat L, Frisken BJ, Bailey AE, De Bruyn, JK. Microrheology and structure of a yield-stress polymer gel. Physical Review E 2006; 73 (4). doi:041405 10.1103/PhysRevE.73.041405.

[19] Kandar AK, Bhattacharya R, Basu JK. Communication: Evidence of dynamic heterogeneity in glassy polymer monolayers from interface microrheology measurements. Journal of Chemical Physics 2010; 133 (7). doi:071102 10.1063/1.3471584.

[20] Gardel ML, Valentine MT, Crocker JC, Bausch AR, Weitz DA. Microrheology of entangled F-actin solutions. Physical Review Letters 2003; 91 (15). doi:158302 10.1103/PhysRevLett.

[21] Valentine MT, Perlman ZE, Gardel ML, Shin JH, Matsudaira P, Mitchison TJ, Weitz DA. Colloid surface chemistry critically affects multiple particle tracking measurements of biomaterials. Biophysical journal 2004; 86 (6):4004-4014. doi:10.1529/biophysj.103.037812.

[22] Crocker JC, Valentine MT, Weeks ER, Gisler T, Kaplan PD, Yodh AG, Weitz DA. Two-point microrheology of inhomogeneous soft materials. Physical Review Letters 2000; 85 (4):888-891.

[23] Levine AJ, Lubensky TC. Two-point microrheology and the electrostatic analogy. Physical Review E 2002; 65 (1). doi:011501 10.1103/PhysRevE.65.011501.

[24] Crocker JC, Hoffman BD. Multiple-particle tracking and two-point microrheology in cells. Cell Mechanics 2007; 83:141-178. doi:10.1016/s0091-679x(07)83007-x.

[25] Sbalzarini IF, Koumoutsakos P. Feature point tracking and trajectory analysis for video imaging in cell biology. Journal of Structural Biology 2005; 151 (2):182-195. doi:10.1016/j.jsb.2005.06.002.

[26] Rogers SS, Waigh TA, Zhao XB, Lu JR. Precise particle tracking against a complicated background: polynomial fitting with Gaussian weight. Phys Biol 2007; 4 (3):220-227. doi:10.1088/1478-3975/4/3/008.

[27] Cheezum MK, Walker WF, Guilford WHQuantitative comparison of algorithms for tracking single fluorescent particles. Biophysical journal 2001; 81 (4):2378-2388.

[28] Savin T. Doyle P.S. Static and dynamic errors in particle tracking microrheology. Biophysical journal 2005; 88 623-638.

[29] Savin T. Doyle P.S. Statistical and sampling issues when using multiple particle tracking. Physical Review E 2007; 76, 021501.

[30] Silburn SA, Saunter CD, Girkin JM, Love GD. Multidepth, multiparticle tracking for active microrheology using a smart camera. Rev Sci Instrum 2011; 82 (3). doi:033712 10.1063/1.3567801.

[31] Keen S, Leach J, Gibson G, Padgett MJ. Comparison of a high-speed camera and a quadrant detector for measuring displacements in optical tweezers. Journal of Optics a-Pure and Applied Optics 2007; 9 (8):S264-S266. doi:10.1088/1464-4258/9/8/s21.

[32] Quan TW, Zeng SQ, Huang ZL. Localization capability and limitation of electron-multiplying charge-coupled, scientific complementary metal-oxide semiconductor, and charge-coupled devices for superresolution imaging. Journal of Biomedical Optics 2010; 15 (6). doi:066005 10.1117/1.3505017.

[33] Carter BC, Shubeita GT, Gross SP. Tracking single particles: a user-friendly quantitative evaluation. Phys Biol 2005; 2 (1):60-72. doi:10.1088/1478-3967/2/1/008.

[34] Czerwinski F, Richardson AC, Oddershede LB. Quantifying Noise in Optical Tweezers by Allan Variance. Optics Express 2009; 17 (15):13255-13269.

[35] Ashkin A. Forces of a single-beam gradient laser trap on a dielectric sphere in the ray optics regime. Biophysical journal 1992; 61 (2):569-582.

[36] Svoboda K, Block SM. Biological applications of optical forces. Annual Review of Biophysics and Biomolecular Structure 1994; 23:247-285. doi:10.1146/annurev.bb.23.060194.001335.

[37] Keen S. High -Speed Video Microscopy in Optical Tweezers. PhD Thesis. University of Glasgow, Glasgow, 2009.

[38] Dufresne ER, Spalding GC, Dearing MT, Sheets SA, Grier DG. Computer-generated holographic optical tweezer arrays. Rev Sci Instrum 2001; 72 (3):1810-1816.

[39] Curtis JE, Koss BA, Grier DG. Dynamic holographic optical tweezers. Optics Communications 2002; 207 (1-6):169-175.

[40] Molloy JE, Padgett MJ. Lights, action: optical tweezers. Contemporary Physics 2002; 43 (4):241-258. doi:10.1080/00107510110116051.

[41] Hemar Y, Pinder DN, Hunter RJ, Singh H, Hebraud P, Horne DS. Monitoring of flocculation and creaming of sodium-caseinate-stabilized emulsions using diffusing-wave spectroscopy. Journal of Colloid and Interface Science 2003; 264 (2):502-508. doi:10.1016/s0021-9797(03)00453-3.

Chapter 2

Heliogeophysical Aspects of Rheology: New Technologies and Horizons of Preventive Medicine

Trofimov Alexander and Sevostyanova Evgeniya

Additional information is available at the end of the chapter

1. Introduction

Geoecological and, first of all, cosmo-heliogeophysical factors exert a powerful regulatory influence on human vital activity [2]. Optimal level of blood circulation is of primary importance in the processes of human adaptation to changes of heliogeophysical environment. Cardio-vascular system is one of the first systems that are included in cascade of adaptive organism reactions. The normal functioning of this system provides rate of blood circulation and consequently tissue metabolism, which are optimal for certain conditions. There are studies that consider the effect of heliogeophysical factors: solar activity and disturbances of the geomagnetic field during the study on the state of the cardio-vascular system [8-13]. The possibility of influence of heliogeophysical factors in early human ontogenesis on the risk of development of cardiovascular diseases was revealed. Convincing data, demonstrating the dependence of the cardiovascular system functioning on changing heliogeophysical environment at different stages of ontogenesis were obtained [14,15]. However, despite the important results, most conducted to date research is devoted to cardiac function and central hemodynamics. At the same time it is known that the blood circulation in an organism is determined by the cardiac function, the state of the vascular bed and the rheological properties of blood. Rheological parameters have a significant impact on the volume and the linear blood flow, determining the values of total peripheral resistance and cardiac output of the blood circulation [16,17]. There are not many publications assessing rheological and hemostatic parameters of blood, which provide optimal rate of blood circulation at changes of heliogeophysical environment [1,18-20]. Thus, in the studies of laboratory of helioclimatopathology in 2008-2009 it was found that the dependence of many rheological (blood viscosity), hemostatic (platelet aggregation, clotting time, bleeding time, prothrombin index) and hemodynamic (arterial pressure, pulse wave velocity, endothelial function) parameters varies with age. A significant age-related dynamics of associations of the physiological parameters with heliogeophysical factors in the prenatal ontogenesis of not only the persons surveyed but also of their parents was marked. It is shown the existence of genetically and epigenetically transferred from generation to generation "relay-race" of individual-generic sensitivity (steadiness) to the influence of various heliogeophysical factors, including galactic and solar flows of protons and electrons of different energies [4].

It is shown that disturbances of blood fluidity with blood hyperviscosity and hypercoagulation are one of the leading links in pathogenesis of chronic cardiovascular diseases (coronary heart disease, hypertension) and also their serious complications [21-23, 5-7]. Increase in blood viscosity and its aggregation potential increase hemodynamic disturbances at cardiovascular diseases and may promote myocardium and vascular remodeling, slowing of neoangiogenesis, endothelium dysfunctions. These processes can be both a consequence and a cause of hypertension [24].

Blood viscosity to a large degree is determined by the erythrocytes aggregation. Increased erythrocytes aggregation leads to occlusion of precapillaries and capillaries by erythrocytes aggregates, slow passage of erythrocytes in the narrow parts of blood channels, the general slowing of peripheral blood circulation. Erythrocytes aggregation is a direct cause of capillaries stasis. As a result of long-termed erythrocytes adhesion oxygen content in the erythrocytes decreases, carbon dioxide removes more difficult – all these processes negatively influence on an organism. Moreover the erythrocytes aggregation is accompanied by their damage with consequent isolation of erythrocytes clotting factors in the blood, which promotes hypercoagulation [22,25].

It is revealed that deformability of erythrocytes is decreased and erythrocytes and platelets aggregation is increased in hypertensive patients. A link between "rigidness" of erythrocytes and mass index of left ventricular myocardium was found. Changes of blood rheological parameters pass ahead disturbances of vasomotor endothelium function. Increase in blood viscosity complicates hemodynamic disturbances in hypertension and can promote myocardium and vascular remodeling [26].

Dependence of functional indices of the blood system on heliogeophysical disturbances is observed in patients with cardio-vascular pathology [27,18,28,29,10]. As an example, the studies, carried out on patients with coronary artery disease, show that at day of the beginning of geomagnetic disturbance, a sludge-phenomenon, perivascular changes, slowing of capillary circulation up to stasis are occurring [18].

It is shown that hypertensive crisis, attacks of angina pectoris, acute conditions of coronary artery disease and cerebrovascular disturbances may be a consequence of the changes of the hemodynamic, functional indices of the blood system, occurring at geomagnetic disturbance [28,10].

All the above defines the importance of the study of complex dependencies of the blood rheological, hemostatic and biochemical indices on changes of heliogeophysical environment and the development of preventive measures in relation to patients with high heliomagnetosensitivity.

The aim. To develop and test elements of a preventive system, reducing the risk of hemodynamic disturbances, crises and their complications in patients with arterial hypertension and high heliomagnetosensitivity of hemorheologic and cardiovascular parameters.

2. Objectives

1. To study the dependence of the rheological and hemodynamic parameters in patients with arterial hypertension on heliogeophysical factors at different stages of ontogenesis.
2. To study the influence of heliogeophysical factors on hemorheological parameters in patients with arterial hypertension in conditions of short-term geomagnetic deprivation of the blood samples.
3. To develop a basis of prevention of heliomagnetotropic reactions in patients with arterial hypertension with the use of light-water-mediated information holographic impacts (according to the patents RF № 2239860 from 05.05.2003 and № 2342149 from 27.12.2008).

3. The contingent and methods

- Patients with arterial hypertension (n = 240), men and women aged 38 to 64 years.
- Blood samples (n = 240) of these patients for the study of rheology (blood viscosity) and hemostatic (clotting time, prothrombin index, bleeding time) parameters.
- Methods of hemodynamic studies (blood pressure, heart rate, pulse wave velocity and endothelial function by the device "Tonocard").
- Hospital Anxiety and Depression Scale HADS.
- The computer program "Helios" to assess the functional dependence of the human systems on heliogeophysical situation in early ontogenesis (Certificate of state registration № 970125 from 24.05.1997). The program "Helios" contains database of 100 years in depth of the daily dynamics of the number and area of sunspots, solar radio emission in biotropic range of 220 mHz and induction of the geomagnetic field. After introduction in a computer information of date of patients` birth, the program algorithm allows to consider the distribution and value of the above mentioned factors in the different periods of prenatal organism development and to distinguish the system with the most heliomagnetosensitivity. In the present study the program has been used to estimate the dependence of hemorheological parameters on prenatal heliogeophysical situation.
- Data of satellite cosmophysical monitoring.
- Computer gaze-discharged visualization (GDV) (by K.G. Korotkov, 2002).

- The installation, shielding the Earth's magnetic field (weakening of the full vector of geo-magnetic induction more than 500 times) structured by Y. Zaitsev (Figure 1).

Figure 1. The installation, shielding the Earth's magnetic field structured by Y. Zaitsev

Figure 2. The screening device "TRODR-1" for light-holographic treatment of drinking water in the weakening geo-magnetic field (A. Trofimov, G. Druzhinin, 2011)

The portable screening device "TRODR-1" (by Trofimov A., Druzhinin G.) for light-holographic treatment of drinking water in the weakening of the total geomagnetic field vector in 300 times (Figure 2).

To determine the functional state of the cardio-vascular system the device "Tonocard" (Russia, Moscow) (Figure 3) was used. The device has Protocol № 14 П-09-24-044 of medical equipment, issued by 23.09.2009 by Federal Service on Surveillance in Healthcare and Social Development of Russian Federation.

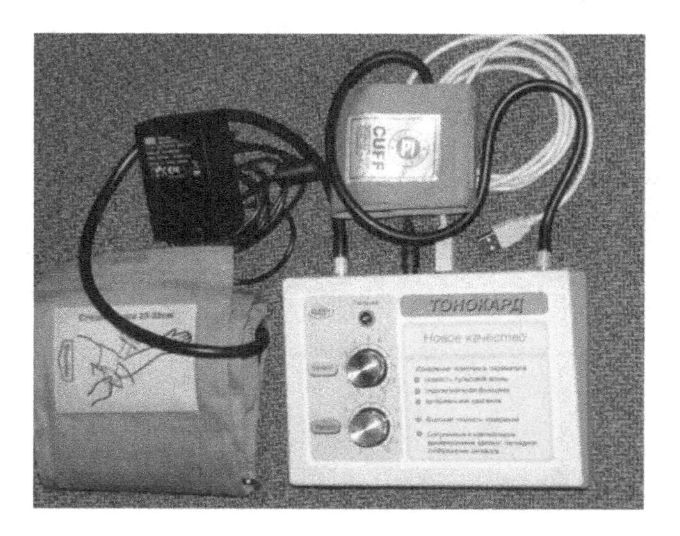

Figure 3. The device "Tonocard"

The following indices were evaluated: systolic and diastolic blood pressure, heart rate, velocity of pulse wave, endothelial function. Endothelial function was determined by the difference in pulse wave velocity before and after short-term (1 min.) occlusion of the brachial artery.

Fractional composition of blood lipoproteins was determined by the method of small-angled X-ray scattering (SAXS), which is analogous to X-ray diffraction analysis, by small-angle X-ray diffractometer Siemens (Germany) in the blood samples of the tested volunteers, physicians with hypertension. The study was conducted at 5 time points before and after experimental course impacts: background (1), the control drinking water (2), holographically treated water (3), the distant impact of the illuminated control hologram (4) and the information capacious hologram (5). In test series (3 and 5) the hologram containing analog information about the complex helioprotective drugs was used.

Heliomagnetoprotective means on water and holographic basis. There is a method of creating a hologram, including separation of the laser beam into the reference and object branches so that the geometric length difference between the branches did not be more than length of the coherent laser source. The disadvantage of this method is that it is not allowed to enter into the hologram nonvizualized physiologically meaningful information. ISRICA together with "Holoart" created and patented (patent RF № 2239860 from 15.11.2004) a new type of hologram recording that uses water-mineral carrier of nonvizualised physiologically meaningful information about the quantum states of different drugs. The object of the holographic recording can be in a state of active or passive transfer of information, or used as the lens of the light beam. The present study used the holograms, which are the quantum analogues of the complex of antioxidant and anticoagulant agents.

Methods of mathematical (correlation) analysis of physiological and cosmophysical parameters.

4. Results

4.1. Clotting blood system as an indicator of organism heliomagnetosensitivity in patients with hypertension

A relationship between rheological, hemostatic, and other blood parameters and the intensity of heliogeophysical factors was detected. A relative increase of the whole blood viscosity (8.50±0.27 cPs vs. the normal value up to 5 cPs) was revealed in 99% of cardiovascular patients, this increase directly correlating with the number of sun spots on the day of the study ($r=0.61$; $p<0.05$). In cardiovascular patients, high sensitivity of the blood system to heliogeophysical factors was detected: increase of solar activity was associated with a trend to a reduction of blood fluidity and increase to its coagulation (Table 1).

Parameter	Main group (n=37)			Reference group (n=31)		
	CT	PI	BT	CT	PI	BT
SSN	-0,36*	0,46*	-0,01	-0,06	-0,15	-0,04
SR	-0,07	0,53*	0,14	-0,18	0,03	-0,08
Ap	-0,41*	-0,41*	-0,31	-0,14	0,02	0,32
Protons	0,06	-0,26	-0,11	-0,08	0,01	0,38*
Electrons	0,38*	-0,18	0,15	0,13	-0,09	0,33

Note: CT – clotting time; PI – prothrombin index; BT – bleeding time;

SSN – sun spots number; SR – solar radiation in the 220 mHz band; Ap – mean circadian geomagnetic index; protons – solar particles with energy > 1meV; electrons – solar particles with energy >0,6 meV

Notes: * significant coefficients of correlation (Spearman, p<0,05)

Table 1. Correlations between hemostasis parameters and heliogeophysical factors during study

No appreciable relationship of this kind was observed in the reference group (n=31). In our view, the obtained data indicate important for pathophysiology and clinic tendency to decrease in blood fluidity at increase in solar activity and reflect the particular sensitivity of the blood to heliogeophysical factors at cardio-vascular diseases.

It is known that one of the basic elements, determining the blood viscosity and characterizing blood system as a whole is the functional state of erythrocytes. For its indirect assessment the erythrocytes sedimentation rate (ESR) was determined. A significant inverse correlation between ESR and area of sun spots ($r=-0.33$; $p<0.05$) on the day of the study was detected in cardiovascular patients ($n=36$); in other words, ESR decreased with increase of solar activity. No relationship of this kind was noted in the reference group ($n=19$).

The spectral and frequency characteristics of the blood, recorded by computer GDV, are sensitive indicators of helio-biospheric effects. A direct correlation ($r=0.51$; $p<0.05$) between the area of GDV fluorescence of blood samples from hypertensive and coronary patients ($n=65$) and the intensity of solar activity, determined by the intensity of stream of protons with energy >10 meV, and a significant inverse relationship ($p<0.05$) of these GDV parameters with the intensity of streams of solar electrons ($r=-0.35$), low-energy (1-10 meV) protons ($r=-0.54$), and geomagnetic activity, evaluated by Ap index ($r=-0.38$), were detected. No relationships of this kind were detected in the reference group ($n=20$).

The presented data confirm the functional dependence of the hemorheological parameters on heliogeophysical environment at the moment of the study that is expressed differently in sick and healthy persons.

Operational internet information (web.site www.sec.noaa.gov) is an important new element of the prognosis of hemorheological dynamics in heliomagnetosensitive patients.

4.2. Dynamics of the rheological parameters of blood (*in vitro*) and cosmophysical interfaces in short-term geomagnetic deprivation

The role of the Earth's magnetic field to maintain hemostasis was determined in the modeled reducing the total vector of geomagnetic induction in 500 times – in conditions of hypogeomagnetic installation (HGMI). In an in vitro study, we found different variants of the reaction of the rheological parameters of blood to the weakening of the geomagnetic field: in the majority of cases (61.5%) there was a decrease in blood viscosity, in 34.6% - increase in blood viscosity and no response – 3.9%, i.e., decrease in blood viscosity with the weakening of the geomagnetic field is found in almost 2 times more than its increase. In conditions of the weakened geomagnetic field statistically significant decrease in the blood viscosity was revealed in patients with hypertension (in background conditions – 8.46 ± 0.26 cPs, in HGMI – 8.05 ± 0.21 cPs, $p = 0.01$).

In the control group there was a trend to a decrease in viscosity in HGMI, but there were no statistically significant differences.

Under experimental conditions of geomagnetic shielding, simulating individual elements of multilevel shielded megalopolis space, essentially modulating the degree of biotropic effects of natural physical factors, the group of cardiological patients exhibited reduction of correlations between blood viscosity and ESR (in vitro) and Wolf numbers (Figure 4).

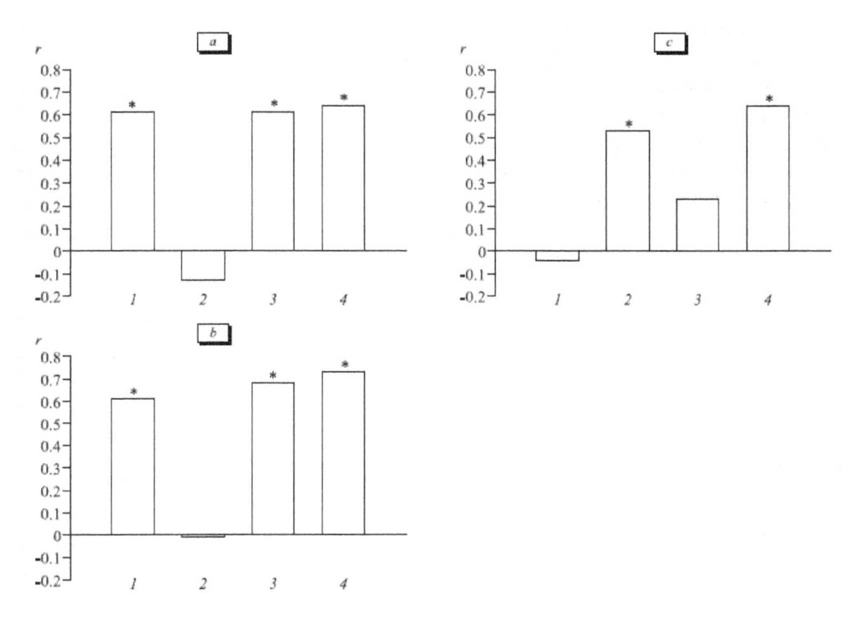

Figure 4. Dynamics of correlations between blood viscosity of cardiovascular patients and solar activity parameters in exposure of blood samples *in vitro* in attenuated GMF. a) number of sun spots; b) area of sun spots; c) protons (solar particles with >1 meV energy). *Significant coefficients of correlation (after Spearman; p <0.05.). 1) main group (basal values); 2) main group (hypogeomagnetic exposure); 3) reference group (basal values); 4) reference group (hypogeomagnetic exposure).

Short-term (30 min) exposure of blood samples from cardiovascular patients to attenuated geomagnetic field led to a significant (9-fold) decrease in the number of significant relationships between blood GDV values and heliogeophysical factors. The only significant inverse correlation was detected: between the area of GDV fluorescence and intensity of protons with >1 meV energy (r=-0.38; p<0.05; n=65). No effects of this kind were detected in the reference group.

Thus, it was found that short-term 30-minute geomagnetic shielding leads to a significant decrease in blood viscosity and reduce its dependence on the intensity of solar and cosmic radiation. In the conditions of the hypogeomagnetic field weakening of associations of the blood system in patients with hypertension with the basic natural regulatory factors: solar activity and cosmic radiation appears. This can have mixed consequences for the functioning of the organism. In general, these data confirm the role of the geomagnetic field in maintaining of rheological homeostasis. Information concerning the level of geomagnetic induction in the analysis of coagulation, blood transfusions, surgery, determining the dose of anticoagulants, etc. should be evidence-based geo-ecological element of hemorheological monitoring of heliomagnetosensitive patients.

We evaluated possible relationship between rheological, hemostatic, and other parameters of the blood system and the heliogeophysical environment status not only during postnatal development, but also at the early stages of ontogenesis. The results indicated a significant association (inverse correlation) between cardiovascular patients' blood viscosity and geomagnetic induction values during the 1st postnatal month (r=0.50; p<0.05; n=32). Significant (p<0.01) inverse correlations between blood viscosity and Sun radiation intensity during the early ontogenesis were detected in 12 subjects of the reference group. Numerous significant coefficients of correlations (p<0.05), indicating an association of coagulation and fibrinolysis processes in the blood of hypertensive and coronary patients with prenatal heliogeophysical fluctuations (changes in solar radiation intensity and geomagnetic induction fluctuations) were detected throughout all periods of intrauterine development (Table 2).

Parameter	**Main group (n=37)**			**Reference group (n=31)**		
	CT	PI	BT	CT	PI	BT
Ap DC	0,13	0,44	-0,04	0,26	-0,08	0,005
Ap 2	-0,02	0,10	0,46*	0,12	0,07	0,10
Ap 3	0,05	0,10	0,06	-0,13	-0,38*	0,42*
Ap 7	-0,38*	0,26	0,35*	-0,23	-0,21	0,32
Ap 8	-0,01	0,37*	0,25	-0,07	-0,08	0,15
Ap 10	-0,01	-0,01	0,50*	0,16	0,06	0,19
SSN 1	-0,15	0,40*	0,27	0,20	0,14	0,16
SSN 2	-0,10	0,38*	0,27	0,11	0,09	0,12
SSN 3	-0,09	0,28	0,40*	0,13	0,02	0,09
SSN 4	-0,15	0,28	0,37*	0,22	0,07	0,14
SSN 8	-0,23	0,37	0,39*	0,05	-0,02	0,10
SSN 9	-0,20	0,28	0,41*	0,16	0,17	0,05
SR DC	-0,52*	-0,08	0,28	-0,003	-0,02	0,28
SR 1	-0,52*	0,26	0,35	0,16	0,18	0,20
SR 4	-0,45*	0,10	0,47*	0,20	0,09	0,17
SR 5	-0,57*	0,04	0,49*	0,09	0,13	0,17
SR 6	-0,58*	-0,01	0,49*	0,25	0,18	0,08
SR 7	-0,63*	-0,01	0,44*	0,14	0,22	0,04

Notes:* significant coefficients of correlation (Spearman, p<0,05)

Ap – mean circadian geomagnetic index;

SSN-sun spots number;

SR – solar radiation in the 220 mHz band;

Table 2. Type of correlations between hemostasis parameters and heliogeophysical factors during different stages (months) of embryogenesis

Short-term attenuation of geomagnetic field led to repeated manifestation of inverse correlation between blood viscosity and intensity of geomagnetic induction during month 3 of early ontogenesis ($r=-0.56$; $p<0.05$; $n=32$) and a direct correlation between ESR and magnetospheric turbulence during the same prenatal period ($r=0.38$; $p<0.05$; $n=36$). Month 3 of embryonal development, characterized by the appearance of the bone marrow hemopoietic function, can be considered as one of the "critical" periods for the formation of functional relationships in the blood system, determining its sensitivity to many exogenous factors, including the heliogeophysical ones.

Comparison of blood samples' GDV parameters before and after short-term exposure to attenuated geomagnetic field showed that these parameters of cardiovascular patients were also associated with prenatal heliogeophysical situation: they exhibited a significant direct correlation ($p<0.05$; $n=65$) with the values of geomagnetic induction during months 1 and 2 of gestation ($r=0.58$, $r=0.69$, respectively) and solar radiation in the 220 mHz band during month 8 of intrauterine development ($r=0.71$) and during birth ($r=0.66$). No associations of this kind were observed under conditions of basal GMF (~49 000 nT at the latitude of Novosibirsk) in any of the groups of patients.

The mechanisms of "heliogeophysical imprinting", discovered by Novosibirsk scientists 20 years ago [30], remain little studied. Involvement of the blood system in imprinting of prenatal environmental factors was detected in cardiovascular patients, which seems important for further studies of mechanisms, including the genetic ones [31], through which the memory about conditions of early development can be imprinted in the cells till adult status and be needed by adult body [32]. Imprinting of heliogeophysical environmental factors, presumably realized through DNA methylation and blocking of gene expression at the earliest stages of prenatal ontogeny [33], can be an important factor in the pathogenesis of cardiovascular and other diseases.

Hence, the dynamics of bioheliogeophysical conjunctions between some human blood parameters under conditions of simulated hypomagnetic space demonstrated its biotropic effect on the blood clotting system in cardiovascular patients: many blood values in these patients proved to be much more functionally dependent on the values of heliogeophysical factors than in patients without hypertension or coronary diseases during the study and during prenatal development.

The results of the work of the section identified the need of creation a system for forecasting the risk of hemodynamic disturbances for the prevention of complications of cardiovascular diseases in the context of geo-ecological factors at different stages of ontogenesis.

4.3. Dynamics of sensitivity in vitro of hemorheological parameters in patients with hypertension to the effects of heliogeophysical factors and holographic information

In 2 series of study it was assessed the dynamics of correlation dependence of the blood viscosity on different stages of ontogenesis in remote effects on the blood tube laser beam (control) and the laser beam, passing through "helioprotective enlightened hologram" with analogous information about antioxidants and anticoagulants.

In the 1-series viscosity parameters in the control (no additional action) had no significant relationships with the intensity of the proton-electron flows at the time of the study. As the effect of the laser beam a significant direct dependence of viscosity on proton flows with energy about 10 meV was revealed. In passing the laser beam through the protective hologram, this dependence was completely leveled (Table 3).

	Pr>1meV	Pr>10 meV	Pr>100 meV	El>0,6 meV	El>2 meV
Background	0,03	-0,11	-0,29	0,37	0,42
1 (control) test	-0,32	0,60*	0,21	-0,51	-0,59*
2 (experiment) test	-0,27	0,04	-0,43	-0,58*	-0,52

Notes:* significant coefficients of correlation (Spearman, $p<0,05$)

Table 3. Correlations of blood viscosity (in tests with holograms in the 1-series of study) from electron-proton components of cosmic rays

In the 2nd series of studies the earlier findings were confirmed: at the impact of the laser a direct dependence on the proton flow with energies of 10 meV occurs, helioprotective information containing in the hologram inverted a character of the relationship, inverse correlation appears (Table 4).

	Pr>1meV	Pr>10 meV	Pr>100 meV	El>0,6 meV	El>2 meV
Background viscosity	-0,34	-0,08	0,11	0,24	0,36
1 (control) test	0,03	0,75*	0,55	-0,10	-0,23
2 (experiment) test	0,18	-0,27	-0,22	0,02	0,18

Notes:* significant coefficients of correlation (Spearman, $p<0,05$)

Table 4. Correlations of blood viscosity (in tests with holograms in the 2-series of study) from electron-proton components of cosmic rays

Thus, the laser induces direct dependence of the blood system, its rheological parameters on solar proton flows, the helioprotective hologram eliminates this dependence.

Because the study used the blood of patients with hypertension, it should be borne in mind with intravenous laser therapy to prevent blood hypercoagulation and thrombotic complications during periods of increased solar corpuscular activity.

Analysis of sensitivity of the blood system (by its viscosity) to the information contained in the holograms in a hypothetical appeal to the "memory" of an organism about helio-geo-physical events in the prenatal ontogenesis was carried out.

From the data presented in Tables 5, 6, it follows that the laser effects on the blood (in vitro) is a stress factor that forces the blood system to access the prenatally shaped experience of interactions with heliogeophysical environment. Helioprotective information contained in the hologram dampens stress, showing another safeguard quality - stress-protective.=

	Geo -1	**Geo Con.**	**Geo 1**	**Geo 2**	**Geo 3**	**Geo 4**	**Geo 5**	**Geo 6**	**Geo 7**	**Geo 8**	**Geo 9**	**Geo 10**	**Geo birth**	**Geo +1**	**Geo total**
background	-0,02	0,05	0,02	0,17	-0,29	-0,02	0,10	-0,19	-0,14	-0,14	0,07	-0,33	0,07	-0,10	-0,02
1 test	0,53	0,91**	0,74*	0,17	0,29	0,64	0,69	0,50	0,79*	0,74*	0,69	0,43	0,43	0,55	0,74*
2 test	0,38	0,72*	0,53	0,33	0,10	0,48	0,57	0,14	0,50	0,43	0,60	0,12	0,29	0,45	0,57

Notes: significant coefficients of correlation (Spearman)

* - p<0,05;** - p<0,01

Geo -1, con.-induction of geomagnetic field (GMF) in the periods before and at the moment of conception

Geo 1-10- induction of GMF in the 1-10-th months of prenatal development

Geo Birth, +1- induction of GMF on the date of birth and in the 1-st month of postnatal life

Table 5. Correlations of blood viscosity (1 test – control; 2 test – with the use of hologram with illumination) from geomagnetic induction in prenatal period

	Geo -1	**Geo Con.**	**Geo 1**	**Geo 2**	**Geo 3**	**Geo 4**	**Geo 5**	**Geo 6**	**Geo 7**	**Geo 8**	**Geo 9**	**Geo 10**	**Geo birth**	**Geo +1**	**Geo total**
initial	0,02	0,02	0,02	-0,21	0	-0,19	0,02	-0,10	-0,14	-0,14	-0,10	-0,14	0,17	0,05	-0,07
1 test	0,60	0,74*	0,60	0,69	0,55	0,38	0,60	0,64	0,43	0,24	0,41	0,24	-0,26	0	0,57
2 test	0,48	0,52	0,48	0,33	0,36	0,14	0,48	0,41	0,21	0,17	0,26	0,12	0,05	0,17	0,36

Notes: significant coefficients of correlation (Spearman)

* - p<0,05;** - p<0,01

Geo -1, con.-induction of geomagnetic field (GMF) in the periods before and at the moment of conception

Geo 1-10- induction of GMF in the 1-10-th months of prenatal development

Geo Birth, +1- induction of GMF on the date of birth and in the 1-st month of postnatal life

Table 6. Correlations of blood viscosity (1 test – control; 2 test – with the use of hologram with illumination) from indices of solar activity in prenatal period

Overall, the data on the meaning of ultra-weak environmental factors in hemorheological dynamics are presented. In addition, based on the data obtained it is promising during a laser therapy to take in account heliogeophysical situation and to use holographic helioprotective and anticoagulant filters.

4.4. Dynamics of heliomagnetosensitivity of an organism in hypertension by the data of course clinico-physiological tests of water-holographic heliomagnetoprotective means

It was noted that in the conditions of repeated short-term holographic impacts many hemodynamic parameters improved: systolic and diastolic blood pressure decreased, pulse wave velocity decreased, endothelial function improved.

At holographic impact, the inverse correlation dependence of endothelial function on such parameters as solar activity (the number and area of sunspots) and proton component of heliogeophysical environment was revealed, i.e. at high concentrations of protons endothelial function decreased (Table 7).

	SSN	SSA	Ap	Am	Pr5
Background	0,54	0,53	-	-	0,48
Control	-	-	-	-	-
Hologram	-0,45	-0,48	-	-	-0,47

Table 7. Significant ($p < 0.05$) correlations of endothelial function with heliogeophysical factors in the background, control and holographic testing

Attention is drawn to the same type of changes in correlations of hemodynamic indices with the proton component of the different energies in the process of transition measurement from the background to the control and from the background to the test.

Thus, according to data obtained Ph.D. V.Ya. Polyakov, course short-term use of holograms with analog helioprotective information (holographic glasses in 5 minutes, the source of "cold light" at a distance of 1.0 m) leads to a change in the above parameters and their interfaces with heliogeophysical factors in the direction indicating helioprotective-effect [34].

The data are presented, demonstrating the expressed heliomagnetoprotective effect of protracted taking of drinking water treated with the use of light -holographic technologies, which is to improve the health and optimize the connection of endothelial function with heliogeophysical environment.

Holographic treatment of drinking water in hypopogeomagnetic installation by way patented in Russia (Russian patent 2342149 on 27.12.2008) led to the creation of a new protective product-heliomagnetoprotective drinking water. Its trial for 2 weeks on a group of volunteers under the control of the dynamics of lipid profile, which is of great importance to hemorheology, was to prove or disprove the expected heliomagnetoprotective effect in relation to the dependence of the content in the blood of lipoproteins of different density on biotropic heliogeophysical factors.

It was carried out an assessment of the dynamics and dependencies of the distribution of the inhomogeneous electron density of lipoprotein macromolecules (nanoparticles of about 10 ÷ 103 A), of their geometric and the weight invariants in venous and capillary blood of the subjects (n = 4) on the intensity of X-ray and radio emission from the sun, the flow of electrons with energies greater than 0.6 and more than 2.0 meV, neutrons, protons with energies greater than 1.0, more than 10,0 and more than 100.0 meV, as well as the ion plasma temperature, as measured by satellites of Goes, at the holographic testing.

Two subjects (K and L) revealed different dynamics of investigated correlations. In the patient K., when taking holographic water, a direct correlation of high-density lipoprotein on the flow of solar protons increased; in the patient P at the direct holographic impact a direct dependence of low density lipoproteins on proton flow with energy of 100 meV weakened (Table 8).

The subject	Background	Control water	Holographic water	Control	Hologram+
O	0,60	0,56	0,62	0,47	0,38
Л	0,37	0,41	0,63	0,45	0,38
K	0,50	0,50	0,77*	0,37	0,47
P	0,58	0,67*	0,73*	0,77*	0,37

Notes:* significant coefficients of correlation (Spearman, p<0,05)

Table 8. Correlations (Spearman) of different density lipoproteins in the capillary blood with the daily values of the cosmic rays of different energies and charges

The data in the table provided by PhD T.V. Kuznetsova, concern only the parameters of capillary blood, in venous blood this dependence was not observed [34].

Therefore, this study allows to make the conclusion about the sensitivity of the blood system, its rheological characteristics to cosmo-heliogeophysical impacts and that the expressed association of hemorheological parameters with heliogeophysical environment is to a large degree determined by its exposure in the period of prenatal development. It was found that susceptibility to cardiovascular disease depended on the peculiarities of interaction of the adult with cosmo-heliogeophysical factors during its early ontogenesis. The data are presented indicating the possible formation of resistance to the development of cardiovascular pathology in individuals with a certain variant of dependence of the blood viscosity on the heliogeophysical conditions in the early stages of ontogenesis. It was found that the short-term 30-minutes geomagnetic shielding led to a considerable decrease in blood viscosity and its dependence on the intensity of solar and cosmic radiation. These findings not only deepen the fundamental understanding of the solar-biosphere relations but also have practical significance for the development of therapeutic and preventive measures during periods of geomagnetic disturbances while managing patients with chronic cardiovascular pathology.

To decrease organism heliomagnetosensitivity we applied helioprotective holograms, containing information on a number of drugs. The application of helioprotective holograms has led to the optimization of relationship of blood rheological parameters in patients with hypertension with heliogeophysical environment, reduce in dependence of blood viscosity indices on solar and cosmic radiation.

In the short-term repeated holographic exposures associations of endothelial function indices with heliogeophysical environment improved.

The course intake of holographically treated water led to helioprotective effect, which consists in a more favorable, antiatherogenic shift of lipid spectrum with increasing cosmic radiation.

5. Conclusions

Blood coagulation system reflects personal pronounced degree of dependence of hemorheological parameters and the whole organism in patients with arterial hypertension on solar and geomagnetic activity at different stages of ontogenesis.

Heliogeophysical environment is an important environmental risk factor for the development of hemorheological and hemodynamic disturbances in patients with arterial hypertension during periods of solar and geomagnetic perturbations.

Short-term exposure of blood samples in conditions of high-gradient changes of the magnetic field of the Earth in the installation, modulating its weakening more than 500 times, results in significant changes of associations of hemorheological and heliogeophysical parameters and proves the role of the geomagnetic field in the mechanisms of maintenance of the hemorheological constancy.

The short-term effect of holographic information on blood samples in patients with arterial hypertension, transmitted to them through a red helium-neon laser, reveals the early-ontogenetic memory of the organism, altering the expression and vector of dependence of the rheological parameters of blood on heliogeophysical situation in the prenatal period of development, as well as on the proton-electron component of solar (cosmic) rays at the moment of the investigation.

The short-term test light-mediated effect of holograms, containing helio-protective information, reveals its preventive character, since it establishes a significant inverse correlation dependence of endothelial function, pulse wave velocity and other hemodynamic parameters on the magnitude of the geomagnetic induction, which biotropic value was studied at experimental geomagnetic deprivation

The course (2 weeks) receiving by patients with arterial hypertension of drinking water with heliomagnetoprotective properties led to an improvement of well-being of patients, stabilization of hemodynamic parameters and manifestation of their advance (48 hours or more)

correlation with heliogeophysical parameters allowing an organism to adapt to high-gradient fluctuations of the natural electromagnetic environment.

Heliomagnetoprotective means used in the form of drinking waterand light-holografic devices open new perspectives for their practical application in the system of geo-environmental prevention of hemorheological and hemodynamic disturbances in patients with arterial hypertension.

Author details

Trofimov Alexander* and Sevostyanova Evgeniya

*Address all correspondence to: isrica2@rambler.ru

International Scientific Research Institute of Cosmic Anthropoecology, Scientific Center of Clinical and Experimental Medicine of SB RAMS, Novosibirsk, Russia

References

[1] Chizhevskii AL. Biophysical mechanisms of erythrocyte sedimentation reaction. Novosibirsk; 1980.

[2] Kaznacheev VP., Trofimov AV. Essays on the nature of living matter and intelligence on the planet Earth. Problems of cosmoplanetary anthropoecology. Novosibirsk: Nauka, 2004.

[3] Kulikov VYu., Voronin AYu., Gaidul KV., Kolpakov VM. Biotropic characteristics of attenuated geomagnetic field. Novosibirsk; 2005.

[4] Trofimov AV., Sevostyanova EV. Dynamics of blood values in experimental geomagnetic deprivation (in vitro) reflects biotropic effects of natural physical factors during early human ontogeny. Bulletin of Experimental Biology and Medicine 2008; 146(1): 100-103.

[5] Becker RC. The role of blood viscosity in the development and progression of coronary artery disease. Cleve Clin. J.Med. 1993; 60(5): 353-358.

[6] Sweetnam PM., Thomas HF., Yarnell JWG, et al. Fibrinogen, viscosity and the 10-year incidence of ischemic heart disease. The Caerphilly and Speedwell Studies. European Heart Journal 1996; 17: 1814-1820.

[7] Struijcer Boudier HAJ. Microcirculation in hypertension. Eur.Heart J. 1999; I(L): 32-37.

[8] Andronova TI., Deryapa NR., Solomatin AP. Heliometeotropic reactions of a healthy and a sick person. L.: Medicine, 1982.

[9] Hasnulin VI. Hasnulina AV., Sevostyanova EV. Northern cardiometeopathies. Novosibirsk:Creative Union "South-West"; 2004.

[10] Breus TK., Komarov FI., Rapoport SI. The medical effects of geomagnetic storms. Clinical medicine 2005; 3: 4-12.

[11] Polyakov V., Trofimov A. Biorhytmological and clinico-functional features of arterial hypertension under geoecological conditions of the North. Alaska Med. 2007; 49(2): 120-126.

[12] Stoupel E., Assali A., Teplitzky I., Vaknin-Assa H., Abramson E., Israelevich P., Kornowski R. Physical influences on right ventricular infarction and cardiogenic shock in acute myocardial infarction. J.Basic Clin.Physiol.Pharmacol. 2009; 20(1): 81-87.

[13] Trofimov A. Impact of the heliophysical factors on man`s lifespan in the circumpolar regions. Perspectives of intake of helio-geroprotectors. Int.J.Circumpolar Health 2010; 7: 356-360.

[14] Trofimov AV, Polyakov VYa., Devitsin DV., Sevostyanova EV. Phenomenon of heliogeophysical imprinting: new possibilities of predicting, diagnosis and prevention of cardiovascular diseases. In: Heliogeophysical factors and human health: proceedings of the International Scientific and Practical Symposium, 15-16 Nov., 2005, Novosibirsk. Novosibirsk; 2007

[15] Trofimov AV. About the possible influence of helio-geophysical factors in prenatal ontogenesis on duration of human life. New horizons of geroprevention: proceedings of the 6-th European congress of biogerontology, 30 Nov.-3 Dec.2008, Netherlands. Netherlands; 2008.

[16] Levtov VA., Regirer SA., Shadrina NH. Blood rheology. M.: Medicine, 1982.

[17] Ivanov KP. Successes and issues in study of microcirculation. Russian Journal of Physiology by name I.M. Sechenov 1995; 81(6): 1-17.

[18] Gurfincel YuI., Lyubimov VV., Oraevskii VN., Parfenova LM., Yurev AS. Effect of geomagnetic disturbances on capillary blood flow in patients with coronary heart disease. Biophysics 1995; 40(4): 793-799.

[19] Ionova VG., Sazanova EA., Sergeenko NP. Influence of helio-geomagnetic field on hemorheological characteristics of people. Aerospace and Environmental Medicine 2004; 38(2): 33-37.

[20] Sevostyanova EV., Trofomov AV., Kunitsyn VG., Bakhtina IA., Kozhevnikova IN. Heliogeophysical factors influence on the rheological properties of blood in patients with chronic cardio-vascular pathology. Bulletin of SB RAMS 2007; 5(127): 94-99.

[21] Dintenfass L. Hyperviscosity in Hypertension. Sydney; 1981.

[22] Dintenfass L. On changes in aggregation of red cells, blood viscosity and plasma viscosity during normal gestation. Clin. Hemorheol.1982; 2(3): 175-188.

[23] Kitaeva ND., Shabanov VA., Levin GYa., Kostrov VA. Microrheology violation of erythrocytes in hypertensive patients. Cardiology 1991; 31(1): 51-53.

[24] Makolkin VI., Podzolkov VI., Pavlov VI., Samoilenko VV. Microcirculation state in hypertension. Cardiology 2003; 5: 60-67.

[25] Galenok VA., Gostinskaya EV., Dikker VE. Hemorheology in disorders of carbonate metabolism. Novosibirsk: Nauka, 1987.

[26] Shlyakhto VV, Moiseeva OM., Lyasnikova EA., Villevalde SV., Emelyanov IV. Rheological properties of blood and endothelial function in hypertensive patients. Cardiology 2004; 4: 20-23.

[27] Oranskii IE., Ilhamdzhanova DS Caught in the magnetic storms. Tashkent: Medicine, 1990.

[28] Komarov FI., Rapoport SI., Breus TK., Oraevskii VN., Elkis IS. To the problem of solar activity influence on clinically important types of pathology. Clinical Medicine 1995; 4(3):8-13.

[29] Stoupel E., Petrauskiene J., Kaledrene R., Abramson E., Sulkes J. Clinical cosmobiology. The Lithuanian study 1990 – 1992. Int. J. Biometeorol. 1995; 38(4): 204 - 208.

[30] Kaznacheev VP., Deryapa NR., Hasnulin VI., Trofimov AV. About the phenomenon of heliogeophysical imprinting and its importance in formation of the types of human adaptive reactions. Bulletin of SB AMS USSR 1985; 5: 3-7.

[31] Tycko B. Am J.Pathol. 1994; 3: 431-443.

[32] Holliday R. Philos.Transact. Royal Soc. 1997; 352(363): 1793-1797.

[33] Bird A. Welcome Trust Rev. 2000; 9: 47-49.

[34] Trofimov AV., Druzhinin GI. Information hologram: theoretical and practical perspectives for ecology and medicine of XXI century. Krasnoyarsk: Polikor, 2011.

Chapter 3

Rheological Characterisation of Diglycidylether of Bisphenol-A (DGEBA) and Polyurethane (PU) Based Isotropic Conductive Adhesives

R. Durairaj, Lam Wai Man, Kau Chee Leong,
Liew Jian Ping, N. N. Ekere and Lim Seow Pheng

Additional information is available at the end of the chapter

1. Introduction

The electronics industry has been striving to find a suitable replacement for lead-based, Sn-Pb solder paste after introduction of legislation to ban the use of lead in electronic products. Due to the toxicity of lead in electronic products, legislation has been proposed to reduce the use of and even ban lead from electronics. Lead-free solders (Pb-free solders) and isotropic conductive adhesives (ICAs) have been considered as the most promising alternatives of lead-based solder [1-2]. ICAs offer numerous advantages over conventional solder, such as environmental friendliness, low temperature processing conditions, fewer processing steps, low stress on the substrates, and fine pitch interconnect capability. Therefore, ICAs have been used in liquid crystal display (LCD), UHF RFID tag antennas, smart card applications, flip–chip assembly and ball grid array (BGA) applications as a replacement to solder [3].

The ICAs materials consist of two components; a polymer matrix and electrically conductive fillers. Traditionally bisphenol-A based epoxies has been used widely used in the electronic packing industry due to their excellent reliability, good thermal stability and high Young's modulus [4]. As the current trend for miniturisation is set to continue towards flexible electronic components with the aim of integrating into sensors or biocompatible electronic components, bisphenol-A is not suitable for this application due to high Young's modulus, hardness and brittleness. Polyurethane (PU) is seen as promising replacement for bisphenol-A based isotropic conductive adhesives due to well-known mechanical properties and can exhibit greater flexibility [4].

Rheological characterisation of pastes materials is the key to understanding the fundamental nature of the ICA suspensions; for example the effect of particle size distributions of silver flakes or powders on the flow and deformation behavior of the pastes. Paste materials are dense suspensions, which exhibit complex flow behavior under the influence of stress. The formulation of new materials such as Polyurethane (PU) based ICAs will require an extensive understanding of the rheological behavior, which is significant for the assembly of flexible electronic devices. A number of studies have reported the rheological behavior of the Diglycidylether of bisphenol-A (DGEBA) based isotropic conductive adhesives with silver flakes as the conventional filler materials [5-7]. But the rheological studies on PU based conductive adhesives are limited. The aim of this study is to investigate the rheological behaviour of concentrated PU and DGEBA based isotropic conductive adhesives. The rheological responses under oscillatory shear stress were examined as a function linear visco-elastic region (LVER), volume fraction and particles size (silver flakes, silver powder and mixture of silver flakes and silver powder).

2. Introduction to Electrical Conductive Adhesives (ECAs)

Electrical conductive adhesives (ECAs) are gaining great interest as potential solder replacements in microelectronics assemblies. Basically, there are two types of ECAs, isotropic conductive adhesive (ICA) and anisotropic conductive adhesive (ACA) (Gilleo, 1995). Although the concepts of these materials are different, both materials are composite materials consisting of a polymer matrix containing conductive fillers. Typically, ICAs contain conductive filler concentrations between 60 and 80 wt.%, and the adhesives are conductive in all directions. ICAs are primarily utilized in hybrid applications and surface mount technology [8]. In ACAs, the volume fractions of conductive fillers are normally between 5 and 10 wt.% and the electrical conduction is generally built only in the pressurization direction during curing. ACA technology is very suitable for fine pitch technology and is principally used for flat panel display applications, flip chips and fine pitch surface mount devices [9]. Compared to conventional solder interconnection technology, conductive adhesives are believed to have the following advantages [10]:

a. More environmental friendly than lead-based solder;

b. Lower processing temperature requirements;

c. Finer pitch capability (ACAs);

d. Higher flexibility and greater fatigue resistance than solder;

e. Simpler processing (no need to use of flux);

f. Non-solderable (inexpensive) substrates can be used (e.g., glass).

Despite the advantages of ECA technology, the replacement of solder by this technology has not been widely adopted by the electronics industry. Lower electrical conductivity than solder [11], poor impact resistance and long-term electrical and mechanical stability [12] are several critical concerns that have limited wider applications of electrically conductive adhesive technology. Numerous studies are being conducted to develop a better understanding of the mechanisms underlying these problems and to improve the performance of conductive adhesives for electronic applications.

In general, there are two conductive pathways for isotropic conductive adhesives. One is genuine conduction, caused by particle to particle contact within the polymer matrix. The other is percolation, which involves electron transport brought about by quantum-mechanical electron tunneling between particles close enough to allow dielectric breakdown of the matrix. Researchers has suggested that percolation is the dominant conduction phenomenon in the early stages of conduction, as the applied current polarizes the conductive adhesive system causing the electrical resistance to drop by charge effects [13]. As currents, especially high currents continue to be applied, polarized particles migrate and further combine, and conduction by particle-to particle contact overwhelms percolation and becomes the dominant conduction phenomenon.

Although electrically conductive adhesives have potential usage and various advantages over solder for surface mount technology (SMT) and microelectronics applications, issues and problems still remain to be solved in order to successfully implement ICAs for solder replacement in electronics assemblies. SMT requires short process times, high yield, high component availability, reliable joints for different components, visual inspection of joints, and capability of repair. ECAs will not be a drop-in replacement for solder in the existing surface mount production lines. First, it will not be cost effective to do so. Special component lead plating and board conduction pad metallisations need to be optimized for conductive adhesives. Standard materials, components and assembly equipment for specific applications need to be developed combining the material vendors", research organizations", and application companies" efforts together. Mechanical bonding strength and electrical conductivity cannot be compromised for the new material development. Fine pitch and thinner lead trends have improved both the pick and placement machine accuracy and the stencil printing process (the laser etched or electroplated stencils and precise stencil printing machine). ICAs have more rigid process requirements for positioning due to their non-selective wetting and lack of self- alignment. Currently major concerns for using ICAs for SMT are the limited availability of components and substrates designed for adhesives, and the lack of methods to predict life-time reliabilities and their relationship to the accelerated life time tests performed as solder joints. Different electrical and mechanical failure mechanisms require one to monitor these properties separately during life-time tests. There are difficulties to inspect the adhesive joints and judge the quality of the joints from visual and x-ray inspection methods, which work for solder joints perfectly. Repairability and reworkability of adhesive joints need to be investigated and improved.

3. Introduction to oscillatory shear testing

Viscoelasticity is the property of materials that exhibit both viscous and elastic characteristics when undergoing deformation. Viscous materials, like honey, resist shear flow and strain linearly with time when a stress is applied. Elastic materials strain instantaneously when stretched and just as quickly return to their original state once the stress is removed. Viscoelastic materials have elements of both of these properties and, as such, exhibit time dependent strain. Whereas elasticity is usually the result of bond stretching along crystallographic planes in an ordered solid, viscoelasticity is the result of the diffusion of atoms or molecules inside of an amorphous material [14].

Before making detailed dynamic measurements to probe the sample's microstructure, the linear visco-elastic region (LVER) must first be defined. This is determined by performing an amplitude sweep test. The LVER can also be used to determine the stability of a suspension. The length of the LVER of the elastic modulus (G') can be used as a measurement of the stability of a sample's structure, since structural properties are best related to elasticity. A sample that has a long LVER is indicative of a well-dispersed and stable system [15]. Therefore, the oscillatory stress sweep is typically used to characterize the visco-elastic effect of emulsions, dispersions, gels, pastes and slurries [16]. A frequency sweep is a particularly useful test as it enables the viscoelastic properties of a sample to be determined as a function of timescale. Within LVER, several segments might have the different visco-elastic properties, therefore, frequency sweep test is performed to study the visco-elastic properties against time [17]. Several parameters can be obtained, such as the Storage Modulus (G') and the Loss Modulus (G").

The oscillatory stress sweep is typically used to characterise the visco-elastic effect of emulsions, dispersions, gels, pastes and slurries. Furthermore oscillatory experiments can be designed to measure the linear or the non-linear visco-elastic properties of dense suspensions such as solder pastes. A sinusoidal stress as a function of the angular velocity (ω) and the stress amplitude (σ_0) is applied on the samples. The applied stress and the resultant strain are expressed as:

$$\sigma = \sigma_0 \sin(\omega.t) \tag{1}$$

$$\gamma = \gamma_0 \sin(\omega.t + \delta) \tag{2}$$

where δ is the phase shift, $\omega=2\pi f$ where f is the frequency, and t is the time. The ratio of the applied shear stress to the maximum strain is called the "complex modulus" (G*) and is a measure of a material's resistance to deformation:

$$G^* = \frac{\tau_0}{\gamma_0} \tag{3}$$

The complex modulus can be divided into elastic and viscous portion representing the magnitude of the strain in-phase and out-of-phase with the applied stress, respectively. The elastic component is called the "storage modulus" and defined as:

$$G' = \left(\frac{\tau_0}{\gamma_0}\right)\cos(\delta) = G^* \cos(\delta) \tag{4}$$

The viscous component or "loss modulus" is defined as:

$$G'' = \left(\frac{\tau_0}{\gamma_0}\right)\sin(\delta) = G^* \sin(\delta) \tag{5}$$

The complex modulus and phase angle can be expressed as functions of the storage and loss modulus:

$$G^* = G' + iG'' \tag{6}$$

In this study the rheological parameters; storage modulus (G′) and loss modulus (G″) is correlated to the solid and liquid characteristic of the DGEBA and PU based isotropic conductive adhesives.

4. Experimental

4.1. Equipment

The rheological curve test measurements were carried out with the Physica MCR 301 controlled stress rheometer. Prior to loading the sample onto the rheometer, the conductive paste was stirred for about 1-2 min to ensure that the paste structure is consistent with the particles being re-distributed into the paste. A sample was loaded on the Peltier plate and the parallel plate was then lowered to the gap of 0.5 mm. The excess paste at the plate edges was carefully trimmed using a plastic spatula. Then the sample was allowed to rest for about 1 min in order to reach the equilibrium state before starting the test. All tests were conducted at 25°C with the temperature controlled by the Peltier-Plate system. Each test was repeated for three times for stabilisation (with fresh samples used for each test).

4.2. Formulation of ICA pastes

In this study, viscosities of formulated isotropic conductive adhesives (ICAs) at different volume fraction of filler with different particles size are investigated. Table 1 show the chemicals used in the formulation of ICAs, which was purchased from Sigma-Aldrich. The epoxy and silver powder/flakes were mixed according to the ratios shown in Table 2. The ICAs materials were formulated into volume fraction (ϕ) of 0.2, 0.4, 0.6 and 0.8. Usually, the filler contents are determined by weight percentage. For example, for the formulation volume fraction of 0.2, 20% of metal filler (silver powder) is mixed with 80% Diglycidylether of bisphenol-A. The summary of all the systems investigated in this study is presented in Table 3. The silver flakes/powder size were measured under scanning electronic microscope (SEM) and found that the flake/ particle size is approximately 10 μm and 250 μm, as shown in Fig. 1 and Fig. 2. An X-ray diffraction test was carried out on the silver flakes and powder; the phases in Fig. 3 show the existence of Ag only, which confirms that the material is pure silver.

Chemical Functions	Chemicals	Manufacturer
Resin	Diglycidylether of bisphenol-A (DGEBA) Polyurethane (PU)	Sigma Aldrich
Curing agents	Ethylene diamine	Merck & Co.
Fillers	Silver flakes and silver powder	Sigma Aldrich

Table 1. Chemicals used in the preparation of isotropic conductive adhesives (ICAs)

Filler size (μm)		Volume fraction of filler
Silver flakes	Silver powder	
10	250	0.2
		0.4
		0.6
		0.8

Table 2. Size and volume fraction of fillers investigated

System	Parameter
S1	0.8-silver flakes/0.2-DGEBA
S2	0.6-silver flakes/0.4-DGEBA
S3	0.4-silver flakes/0.6-DGEBA
S4	0.2-silver flakes/0.8-DGEBA
S5	0.8-silver powder/0.2-DGEBA
S6	0.6-silver powder/0.4-DGEBA
S7	0.4-silver powder/0.6-DGEBA
S8	0.2-silver powder/0.8-DGEBA
S9	0.8-silver flakes+powder/0.2-DGEBA
S10	0.6-silver flakes+powder/0.4-DGEBA
S11	0.4-silver flakes+powder/0.6-DGEBA
S12	0.2-silver flakes+powder/0.8-DGEBA

System	Parameter
S13	0.8-silver flakes/0.2-PU
S14	0.6-silver flakes/0.4-PU
S15	0.4-silver flakes/0.6-PU
S16	0.2-silver flakes/0.8-PU
S17	0.8-silver powder/0.2-PU
S18	0.6-silver powder/0.4-PU
S19	0.4-silver powder/0.6-PU
S20	0.2-silver powder/0.8-PU
S21	0.8-silver flakes+powder/0.2-PU
S22	0.6-silver flakes+powder/0.4-PU
S23	0.4-silver flakes+powder/0.6-PU

Table 3. Summary of the systems investigated in this study

Figure 1. Scanning Electron Microscope (SEM) microstructure of silver flakes

Figure 2. Scanning Electron Microscope (SEM) microstructure of silver flakes

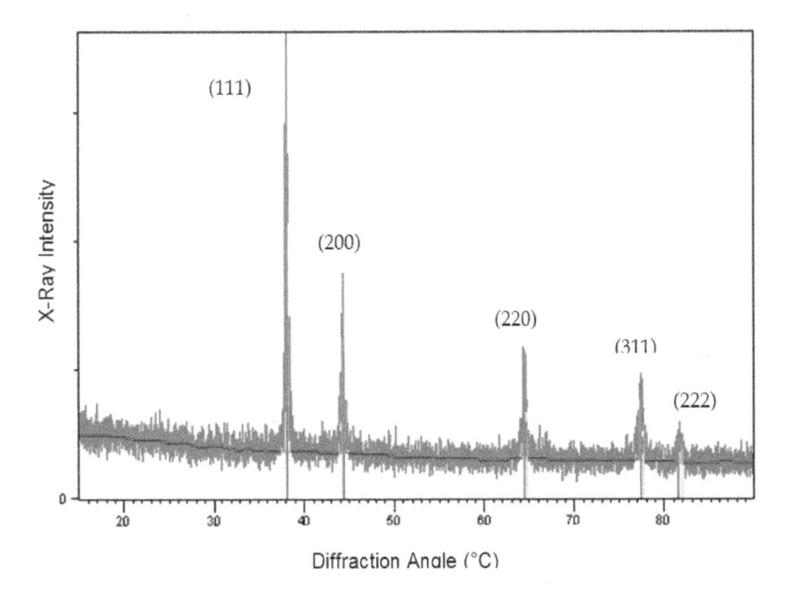

Figure 3. X-ray diffraction pattern for silver (Ag)

4.3. Oscillatory stress sweep test

In the oscillatory stress sweep experiment, initially a large stress sweep range of 0.0001-1000 Pa is applied to all the pastes samples. The oscillatory stress results showed that there are inconsistency in the measured parameters; storage modulus (G′) and loss modulus (G″) at low shear stress. At higher volume fractions, the rheometer had difficulty in taking consistent measurement at shear stresses of 0.001 Pa as opposed to lower volume fractions. This is the reason why some of the rheological data is presented at different shear stresses. This indicates the development of inherent structural strength as a result of the transition the paste undergoes from Newtonian to Non-newtonian, due to the addition of filler materials.

The linear visco-elastic region is defined as the maximum deformation can be applied to the sample without destroying its structure. It should be noted here that the linear data is not particularly relevant for real application processing but can be useful in looking for particle-particle interactions [18]. The length of the LVE region of the elastic modulus (G') with respect to the applied shear stress can be used as a measurement of the stability of a sample's structure, since structural properties are best related to elasticity prior to structural breakdown. In the LVE region, the particles stay in close contact with each other and recover elastically to any applied stress or strain. As a result, the sample acts as a solid and the structure remains intact.

5. Results and discussion

5.1. DGEBA based isotropic conductive adhesives

For the DGEBA epoxy formulation with silver flakes at $\phi = 0.2$, the loss modulus (G″) was greater than the storage modulus (G′), as shown in Fig. 4. The G′ showed a LVE region up to 0.5 Pa after which the G′ values dropped showing a structural breakdown in the paste. The loss modulus, G″ value is constant with increasing shear stress as it gives the response which is exactly out of phase with the imposed perturbation, and this is related to the viscosity of the material.

A similar trend was observed at ϕ =0.4, but with a higher LVE region up to 1 Pa followed by structural breakdown, shown in Fig. 4. At $\phi = 0.6$ and $\phi = 0.8$, the measured storage modulus (G′) is greater than loss modulus (G″) with increasing shear stress. In addition, as the volume fraction is increased from $\phi = 0.2$ to 0.8, the measured LVE region increases from 0.5 Pa, 1 Pa, 10 Pa and 100 Pa, respectively prior to structural breakdown. The shift of LVE region to higher stress range could be due to the strong interaction between different layers of flakes within the system.

Fig. 5 represents the DGEBA epoxy formulated with silver powder with a particle size of 250 μm. At $\phi = 0.2$ and $\phi = 0.4$, the G″ was greater than G′, which indicates the liquid-like behaviour of the paste is predominant, as shown in Fig. 5. For the volume fraction of $\phi = 0.6$ and $\phi = 0.8$, the storage modulus (G′) was greater than loss modulus (G″). At lower volume fraction $\phi = 0.2$ and $\phi = 0.4$, the addition of silver particles did not affect the

Newtonian continuous phase of the epoxy resin. Hence the paste did not show any structural breakdown as observed for silver flakes. The measured LVE region for ϕ = 0.6 and ϕ = 0.8 was up to 0.8 Pa and 1 Pa, which is lower when compared to the DGEBA formulated silver flakes. Beyond the LVE region, the flocculation of silver powder in the DGEBA system is easily broken down over narrow range of shear stress as illustrated by Fig. 5. The results show that a larger particle size has lower contact surface area and has poor dispersion ability.

A bimodal distribution was formulated with a mixture of silver flakes and silver powder, as shown in Fig. 6. As with previous systems, at ϕ = 0.2 and 0.4, the G″ is greater than G′ due to lower concentration of the silver flakes and powder in the systems. However, at ϕ = 0.2 and 0.4, G′ value increases with the applied shear stress and gradually begins to drops after 0.2 Pa. At ϕ = 0.6 and ϕ = 0.8, the LVE region has increased up to 10 Pa and 50 Pa. These values are higher than DGEBA/silver powder system but lower than DEGBA/silver flakes systems. Previous study by Walberger and Mchugh [19] concluded that there will be always an increase in G′ and G″ due to the addition of filler but where the increase in both functions with addition of filler is not the same, the effect on G′ is considerably greater within the linear visco-elastic region. Beyond the LVE region, the paste sample showed a gradual structural breakdown as opposed to silver flakes and powder systems. The results seem to indicate that the flake in the system restricts the movement of the particles, which delays the structural breakdown.

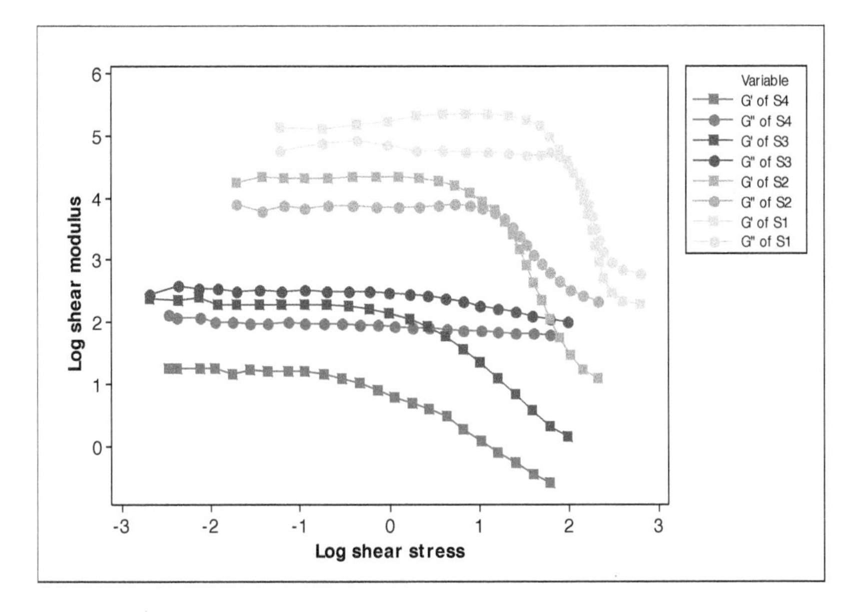

Figure 4. Silver flakes with DGEBA epoxy resin

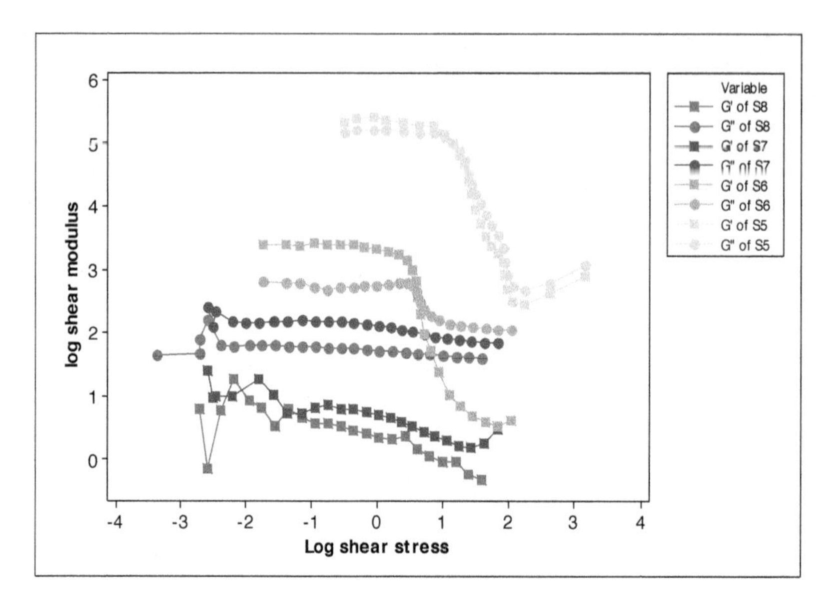

Figure 5. Silver powder and DGEBA epoxy resin

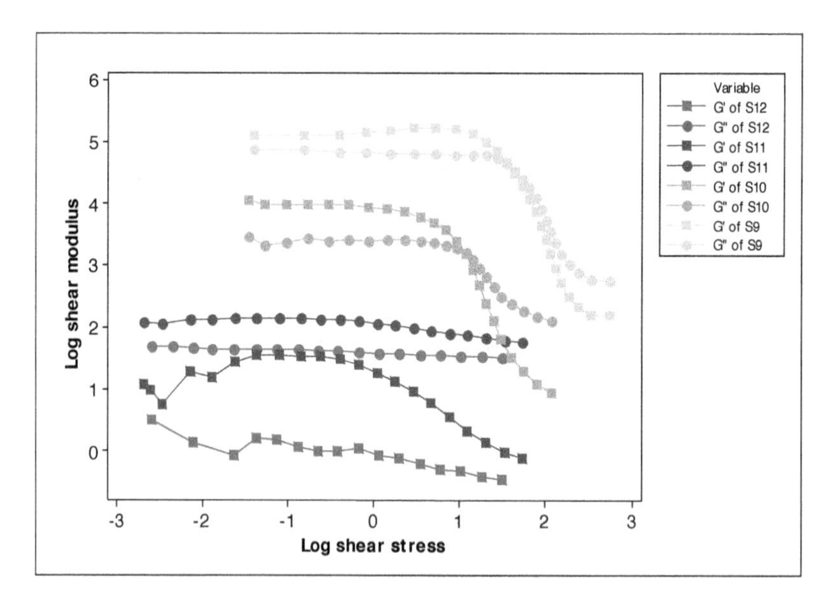

Figure 6. Silver flakes and powder with DGEBA epoxy resin

5.2. PU based isotropic conductive adhesives

The PU systems with silver flakes as the filler material showed that the constant G″ at ϕ = 0.2 and 0.4, while the G′ decreased after 0.01 Pa, as shown in Fig. 7. The LVE region of the PU at ϕ = 0.2 and 0.4 was approximately 0.01 Pa meanwhile ϕ = 0.6 and 0.8 was approximately 5 Pa and 10 Pa, respectively. Beyond the LVR region, a gradual structural breakdown was observed for sample at ϕ = 0.6 and 0.8 as opposed to DGEBA/silver flakes system. But the overall measured solid (G′) and liquid (G″) characteristic of PU based ICAs were lower compared to DGEBA based ICAs, which could prove to be attractive for the assembly of flexible electronic devices.

At lower volume fractions ϕ = 0.2 and 0.4, the G′ did not change significantly with increasing shear stress, as shown in Fig. 8. The observed trend was similar to the DGEBA/silver powder system. When the volume fraction was increased to ϕ = 0.6 and 0.8, the samples showed a LVE range of up to 9 Pa and 20 Pa, respectively. The linear region measured for PU/silver powder was considerably higher than DGEBA/silver flakes systems. Despite the difference in LVE region, both these systems (PU/silver powder and DGEBA/silver flakes) showed a rapid structural breakdown. When the volume fraction ϕ exceeds 0.50 under equilibrium condition with no imposed flow, the silver powder system which is a monodispersed hard sphere suspension begins to order into a macrocrystalline structure of face centered cubic (fcc) or hexagonally close packing (hcp). With increasing applied stress, the drop in G′ and G″ arises forced flow of three dimensionally ordered structures of fcc or hcp. At volume fraction of ϕ = 0.8, an increase in G′ and G″ was observed after 200 Pa and similar result was observed for DGEBA/silver powder formulation. At high volume fraction above 50 % by volume for hard sphere suspensions, the increase in G′ and G″ could be attributed to development of lubrication stress as a result of close network formed between particles. This causes a strong hydrodynamic force; considerable amount of solvent is trapped interior to the particle cluster. The trapping of the solvent apparently decreases the mobile solvent volume fraction, or in effect, increase the particle volume fraction [20].

A bimodal distribution system with silver flakes and silver powder was formulated with PU. In this system, the G′ is greater than G″ for all volume fractions and the measured LVE region up to 0.1 Pa for ϕ = 0.2, 0.4, 0.6 and 0.01 Pa for ϕ = 0.8, as shown in Fig. 9. This system showed the lowest LVE region when compared to all the other systems. This could be due to the orientation of crystal in the direction of closest packing of the spheres is aligned to flow velocity, while the planes containing the closest packing are parallel to the shearing surfaces. Beyond the linear region, the sample showed a gradual breakdown in the paste structure as opposed to silver powder system. Beyond the linear region, the hydrodynamic force prevents them from sustaining their ordered state by forming a three-dimensional network or clustering, which results in the structural breakdown.

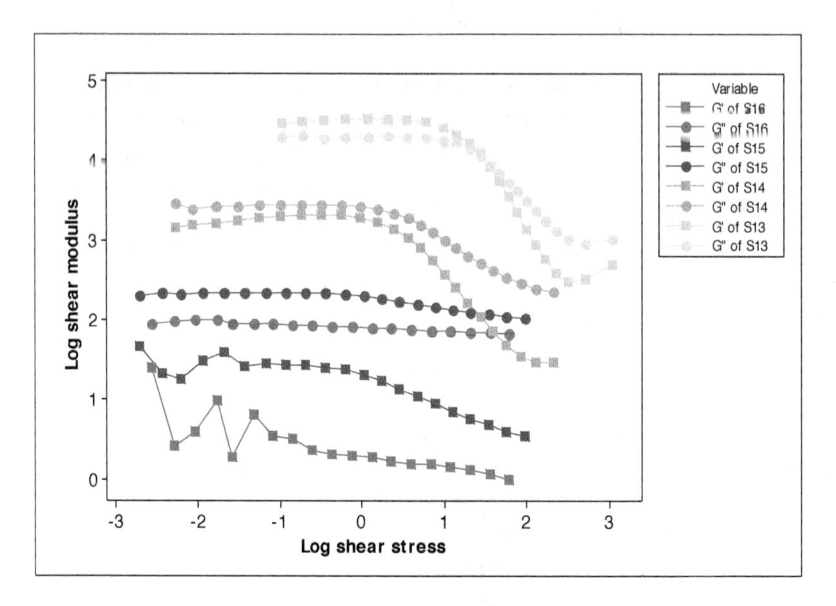

Figure 7. Silver flakes and Polyurethane (PU) resin

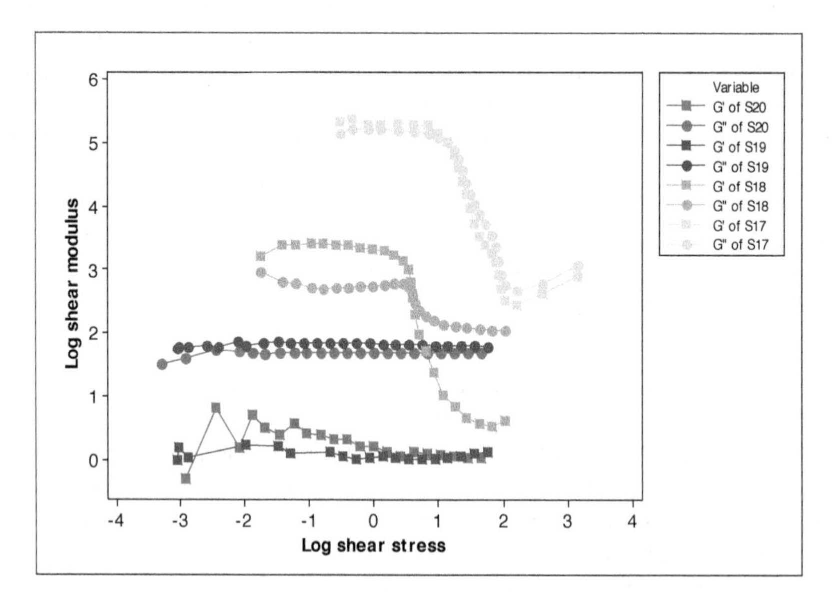

Figure 8. Silver powder and Polyurethane (PU) resin

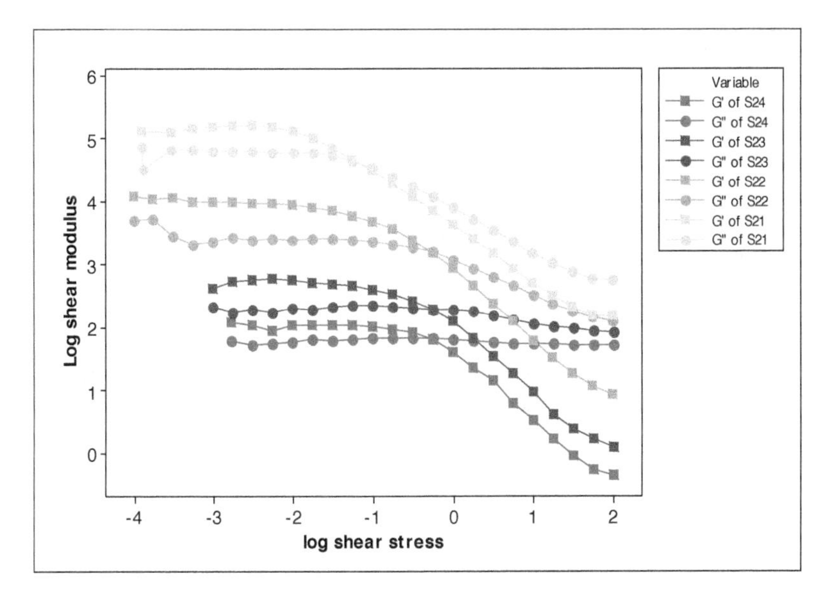

Figure 9. Silver flakes and powder with Polyurethane (PU)

6. Conclusions

From the oscillatory test on PU and DGEBA it is clear that G′ and G″ could be used diagnostically to assess the state of the dispersion, since the linear visco-elastic region varies from one system to another. The study found that the volume fraction of the filler materials is shown to affect the G′ and G″ values. In addition, the particle size of the fillers is found to also influence the flow behaviour of the systems. The study showed that the processability of the systems is related to the solid (G′) and liquid (G″) characteristic of the material beyond the linear visco-elastic region. The extent of the structural breakdown could be used to determine the stability of the formulated systems. The silver powder based PU and DGEBA experience a rapid structural breakdown and increased in G′ and G″ values at higher shear stress. For the PU system, the mixture of silver powder and flakes produced a much stable system with a gradual structural breakdown as opposed to DGEBA systems. In addition, the solid (G′) and liquid (G′) characteristic of PU were lower compared to DGEBA could be a drop in replacement for DGEBA based isotropic conductive adhesives.

Acknowledgements

The authors would like to acknowledge Fundamental Research Grant Scheme (FRGS), Ministry of Higher Education (MoHE) for providing financial support under grant no: FRGS/

2/2010/ST/UTAR/03/3. The author would like to also acknowledge Prof Samjid Mannan at King's College, London, UK.

Author details

R. Durairaj[1], Lam Wai Man[1], Kau Chee Leong[1], Liew Jian Ping[1], N. N. Ekere[2] and Lim Seow Pheng[1]

*Address all correspondence to: rajkumar@utar.edu.my

1 Department of Mechanical and Material Engineering, Faculty of Engineering and Science (FES), Universiti Tunku Abdul Rahman (UTAR), Jalan Genting Kelang, Setapak, Kuala Lumpur, Malaysia

2 School of Technology, University of Wolverhampton, Technology Centre (MI Building), City Campus – South, Wulfruna St, Wolverhampton, United Kingdom

References

[1] J.H. Lau, C.P. Wong, N.C. Lee, S.W. Lee, Electronics manufacturing with lead-free. Halogen-free, and conductive-adhesive materials, McGraw-Hill, 2003.

[2] C.P.Wong and Yi Li, Recent advances of conductive adhesives as a lead-free alternative in electronic packaging: Materials, processing, reliability and applications Journal of Materials Science and Engineering, 2006, 51, 1-35.

[3] M. Irfan and D. Kumar, Recent advances in isotropic conductive adhesives for electronics packaging applications International Journal of Adhesion & Adhesives, 2008, 28, 362-371.

[4] Cheng Yang, Mathew M.F. Yuen, Ba Gao, Yuhui Ma: in Proc of Electronic Component and Technology Conference, USA, 2009, 1337

[5] R. Durairaj R, N.N. Ekere, B. Salam, Thixotropy flow behaviour of solder and conductive adhesives paste J Material Science: Materials in Electronic, 2004; 15, 677-683

[6] R. Durairaj, S. Mallik, A. Seman, A. Marks and N.N. Ekere, Rheological characterisation of solder pastes, isotropic conductive adhesives used for flip chip assembly Journal of Materials and Processing Technology, 2009, 209, 3923.

[7] R. Durairaj, Lam Wai Man and S. Ramesh, Rheological Characterisation and Empirical Modelling of Lead-Free Solder Pastes and Isotropic Conductive Adhesive Pastes Journal of ASTM International, 2010, 7, 7.

[8] Perichaud, M. G., Deletage, J. Y., Fremont, H., Danto, Y., and Faure, C. (2000). Reliability Evaluation of Adhesive Bonded SMT Components in Industrial Applications, 40, (pp.1227-1234), Microelectronics Reliability

[9] Development of Conductive Adhesives Filler with Low-melting-point Alloy Fillers, (pp.7-13), Port Erin, Isle of Man, British Isles: International Symposium on Advanced Packaging Materials, Port Erin, Isle of Man, British Isles

[10] Liu, J. and Lai, Z., "Overview of Conductive Adhesive Joining Technology in Electronics Packaging Applications", 3rd International Conference on Adhesive Joining and Coating Technology in Electronics Manufacturing, pp. 1-17, 1998.

[11] Conductive Adhesives for SMT and Potential Applications, IEEE Transactions on Components, Packaging, and Manufacturing Technology, Part B, 18 (2), 284-291.

[12] Liu, J., Gustafsson, K., Lai, Z., and Li, C., "Surface Characteristics, Reliability, and Failure Mechanisms of Tin/Lead, Copper, and Gold Metallizations", IEEE Transactions on Components, Packaging, and Manufacturing Technology, Part A, vol.20, pp. 21-30, 1997.

[13] Ritter, G. W., "Electrical Current Effects on Conductive Eposies", Proceedings of the 2nd Annual Meeting of the Adhesion Society, pp.56-59, 1999.

[14] Bullard, J. W, Pauli, A. T., Garboczi, E. J., Martys, N. S. (2009). Comparison of viscosity-concentration relationships for emulsions. Journal of Colloid and Interface Science, 330, 186-193.

[15] Durairaj, R., Mallik, S., Seman, A., Marks, A., and Ekere, N. N. (2009). Rheological characterisation of solder pastes, isotropic conductive adhesives used for flip chip assembly. Journal of Materials and Processing Technology, 209, 3923

[16] Mewis, J. and Wagner, N. J. (2009). Current trend in suspension rheology. Journal of Non-Newtonian Fluid Mechanical, 157, 147.

[17] Lapasin, R., Sabrina, P., Vittorio, S., and Donato, C. (1997). Viscoelastic properties of solder pastes. Journal of Electronic Materials, 27, 138-148.

[18] H. A. Barnes, A Review of the Rheology of Filled Viscoelastic Systems, Rheology Reviews 2003, 1 – 36.

[19] J.A. Walberer and A.J. McHugh, "The linear viscoelastic behavior of highly filled polydimethylsiloxane measured in shear and compression", Journal Rheology, 45(1), 2001, pp. 187-201.

[20] Lee Jae-Dong, So Jae-Hyun, and Yang Seung-Man, Rheological behavior and stability of concentrated silica suspensions, Journal Rheology 43 (5), September/October 1999

Chapter 4

Performance of Fresh Portland Cement Pastes – Determination of Some Specific Rheological Parameters

R. Talero, C. Pedrajas and V. Rahhal

Additional information is available at the end of the chapter

1. Introduction

The hard, strong and durable cement–based product required by the user is only achieved following a period of plasticity but the attention paid to its fresh properties is small, despite the far–reaching effects of inadequate fresh performance. Pumping, spreading, moulding and compaction all depend on rheology and thanks to an increasingly scientific approach it is becoming possible to predict fresh properties, design and select materials and model processes to achieve the required performance. Rheology is now seriously considered by users, rather than being seen as an inconvenient and rather specialised branch of cement science.

A number of papers [1-3] have been published on the variations in the technological properties of Portland cement blended with active mineral additions. One of the properties that varies significantly once hydration begins is rheology, with the change of state in the material, in whose measurement and analysis a series of different methodologies are called into play [4-7].

The rheological behaviour of pastes, mortars and concretes continues to be a subject of analysis in light of the large number of factors involved in cement blending, mixing and hydration (such as type of cement, type and proportions of mineral additions and presence or otherwise of admixtures). Very generally, rheology describes the deformation of a "body" subjected to loading. The "body" in this case refers to solid and liquid materials. Ideal solids deform elastically, whereby the energy required in the deformation is recovered when the load is removed. Ideal fluids, by contrast, deform irreversibly because they flow. The energy required for deformation is dissipated in the fluid as heat and cannot be recovered by merely removing the load involved. Real bodies behave neither like ideal solids nor ideal fluids, and may be irreversibly deformed under the effect of sufficiently large forces. The vast majority of liquids exhibit behaviour somewhere in between fluids and solids: they are viscoelastic bodies. Solids may be subjected to perpendicular and tangential stress, whereas fluids can be subjected to tangential (shear) stress only.

Fluid viscosity, represented as η, is defined as its resistance to flow when subjected to shear stress. Newton was the first to formulate a law for viscosity, known as his friction law, according to which for ideal liquids:

$$\tau = \eta \cdot \gamma \quad (1)$$

where τ is shear stress, η is viscosity and γ is strain or the shear rate. The correlation between shear stress, τ, and strain, γ, is defined as the liquid's fluid behaviour This correlation, known as a flow curve, is plotted in Fig. 1.

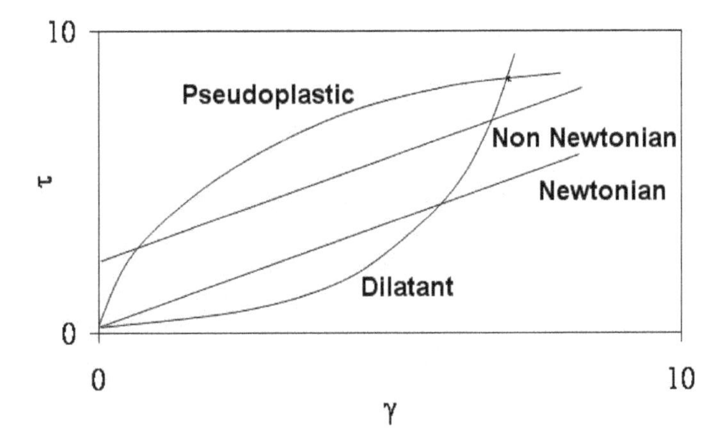

Figure 1. Fluid types.

For ideal liquids, the flow curve is a straight line: the quotient of all the τ-γ pairs on that line is a constant. In other words, viscosity, η, is unaffected by changes in shear rate. All liquids for which this relationship holds are called "Newtonian" and all others "non-Newtonian" fluids. Cement paste fluid behaviour cannot be described by a flow curve such as in Fig. 1. Rather, since these pastes tend to form flocs, a certain minimum shear stress needs to be applied before they begin to flow. In this case the flow equation can be expressed as shown below.

$$\tau = \tau_0 + \mu \cdot \gamma \quad (2)$$

This equation describes the so-called Bingham model, where τ is shear stress; τ_0, threshold shear stress; μ, plastic viscosity; and γ, shear rate or strain. Plastic viscosity is an indication of the number and size of the flocs, while threshold shear stress is a measure of the strength and number of interactions severed when stress is applied.

In addition to the difference between Newtonian and non-Newtonian fluids, the latter may exhibit *dilatant* or *thixotropic* behaviour, as illustrated in Fig.1 [8, 9].

Actually, there are qualitative and quantitative disagreements between the results for cement paste reported by different research workers. The flow curve has been reported to fit several different mathematical forms, all of which indicate the existence of a yield stress:

Bingham [10]

$$\tau = \tau o \ + \mu\dot{\gamma} \tag{3}$$

Herschel–Bulkley [11]

$$\tau = \tau o \ + \ A\dot{\gamma}B \tag{4}$$

Robertson–Stiff [12]

$$\tau = \ A(\dot{\gamma} + \ B)C \tag{5}$$

Modified Bingham [13]

$$\tau = \tau o \ + \mu\dot{\gamma} + \ B\dot{\gamma}2 \tag{6}$$

Casson [11]

$$\tau^{1/v} = \tau o^{1/v} + \ (\mu\dot{\gamma})^{1/v} \tag{7}$$

De Kee [13]

$$\tau = \tau o \ + \mu\dot{\gamma}e - A\dot{\gamma} \tag{8}$$

Yahia and Khayat [13]

$$\tau = \tau_0 + 2\left(\tau_0\mu_p\right)^{1/2}\left(\dot{\gamma}e^{A\dot{\gamma}}\right)^{1/2} \tag{9}$$

where A, B and C are constants.

Additionally the numerical values reported for the rheological parameters cover a very wide range, which cannot be wholly explained by variations in the materials used. It can only be accounted for by accepting that differences in experimental technique and apparatus of different workers have a much greater effect than has been generally realised. Differences in the shear history at the time of test, undetected plug flow and slippage at the smooth surfaces of a viscometer could all combine to give experimental variations as large as those reported. However, there is general agreement on two fundamental qualitative aspects of the behaviour of cement pastes.

First, the material breaks down during the test and hysteresis loops with the downcurve falling to lower stresses than the upcurve are obtained when the flow curve is determined over a short cycle time. The shape changes systematically with increasing cycle time through loops with a crossover point to loops showing structural build up [14], attributable to chemical reaction during the course of the test, but Hattori and Izumi [15] explained the effect in terms of the competition between coagulation and deflocculation processes. The apparent need to fit the range of models in equations 1–7 may be the result of not allowing for the possibility of structural breakdown during the test.

Second, the material has a yield stress which decreases, in line with reductions in the apparent viscosity indicated by the rest of the curve, as the total amount of shearing energy experienced by the paste increases. Thus successive hysteresis loops fall to progressively lower values of torque in a coaxial cylinders viscometer [16], yield stress and plastic viscosity fall to an equilibrium value as the time of mechanical mixing is increased [17] and the effect can be quantified in terms of the total shear energy received by the sample prior to the test [18,19]. This structural breakdown has been amply confirmed by experiments carried out under both continuous steady shear rate and continuous steady stress. In the former the relationship between shear stress and time is affected by the shear rate in the experiment and was explained theoretically by Tattersall [16] using a linkage theory, in which the links between particles are broken by the work done in shearing the paste.

A material's rheology is measured with rheometers or viscometers. A rheometer is a device that measures the viscoelastic properties of solids, semi-solids and fluids, whereas a viscometer is a more sensitive instrument that cannot accommodate large particles.

The two main types of viscometers are:

a. native stress-controlled viscometers, in which the shear stress is user-defined to find the respective shear rate (or strain)

b. native strain- (or shear rate)-controlled viscometers, in which the strain is user-defined to find the respective shear stress. [20].

With the ultimate aim of researching the effect of active and non active (*fillers*) mineral additions on pure Portland cement rheology, the impact of the variables involved on the findings was analysed in this first stage. After establishing a measuring scheme, the variations in paste rheological behaviour were determined from initial hydration up to the first nadir on the calorimetric curve. Its trials were conducted on two pure Portland cements with very different potential mineralogical compositions, whose behaviour in conduction calorimetry and response to sulphate attack had been studied in prior research [21-32]. Finally, this rheological study of PCs is much involved with segregation phenomena and workability of their respective concretes and mortars and, as a consequence, with their durability.

On the other hand and with regard to the filler, the particle surfaces are positively or negatively charged during grinding with Portland clinker and gypsum and/or when mixing process with water and aggregates, consequently and respectively attracting OH- and Ca2+ ions [33] very at the start of the hydration. This first layer of anions or cations in turn attracts a second cluster of Ca2+ or OH- ions, respectively. As the ionic layer thickens, the electrostatic force of the particles declines [34]. Besides this, the following much more important consideration has to be also taken into account when the hydration moves forward: all inorganic particles assume a charge when dispersed in water.

The charged particle surface then attracts a layer of counter-ions (ions of the opposite charge) from the aqueous phase. Due to ionic radio considerations, the strongly adsorbed counter-ions will not fully offset the surface charge. A second layer of more loosely held counter-ions then forms. At a certain distance from the particle surface, the surface charge will be fully balanced by counter-ions. Beyond this point, a bulk suspension with a balance of negative and positive electrolyte exists. The size of the double layer will depend firstly on the amount of charge on the particle surface. A large charge, whether positive or negative, will result in a large double layer that stops particles getting close to each other because of the electrostatic repulsion between those particles carrying the same electrical charge. This situation is typical of stable (deflocculated) suspensions having a low viscosity. Conversely, a low surface charge requires fewer counter-ions and smaller double layers. Accordingly, particles then tend to flocculate which leads to high viscosity suspension. The zeta potential (mV) can be related to the energy needed to shear the particle and its inner layer of counter-ions away from the outer layer / bulk medium. This phenomenon has been illustrated in Fig. 2. In short, as it was mentioned earlier that particle charge influences the double layer size and so the zeta potential. Thus, the potential in this region decays with distance from the surface, some distance until it becomes zero (Fig. 2).

When a voltage is supplied to a solution with dispersed particles, the particles are attracted by the electrode of opposite polarity, together with the fixed layer and part of the diffuse double layer. The potential in the limit of the unit, said cutting plane, between the particle and its ionic atmosphere surrounding medium, is called zeta potential.

The zeta potential is a function of the charged surface of a particle, any adsorbed layer at the interface and the nature and composition of the medium in which the particle is suspended.

The zeta potential can be calculated with the following expression Smoluchowski's:

$$\zeta = \frac{4\pi\eta}{\varepsilon} \times U \times 300 \times 300 \times 1000 \tag{10}$$

ζ=Zeta Potential (mV)

η=Viscosity of Solution

ε+Dielectric Constant

$U = \frac{v}{V/L}$: Electrophoretic Movility

v=Speed of Particle (cm/sec)

V= Voltage (V)

L= The distance of Electorode

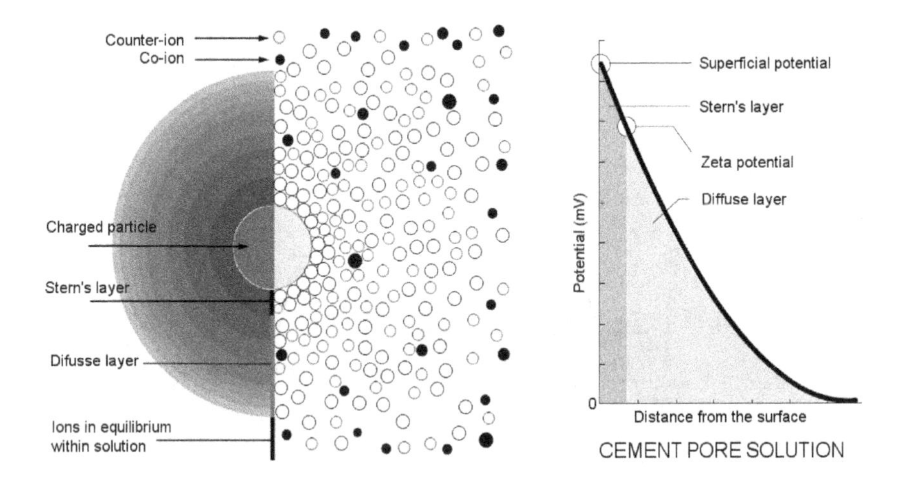

Figure 2. Electrostatic phenomenon in a solution for a charged particle. Graphical description of the Zeta potential.

Determination of the zeta potential is very simple. Applying a controlled electric field by means of electrodes immersed in a sample suspension, this causes the charged particles to move through the electrode of opposite polarity. Viscous forces acting on the particle in motion tend to oppose this movement, establishing a balance between the two forces of electrostatic attraction and viscous drag. the zeta potential has a considerable influence on the rheology of cement, so that increasing the magnitude of the zeta potential (both negative and positive) increases the low shear viscosity. The zeta potential depends on conditions of temperature, pH and others.

2. Materials and methodology

To ascertain the differences in the response depending on the variations in the rheological trial parameters, pastes were prepared with ordinary type CEM I Portland cement [35] (whose mineralogical composition and other physical parameters, midway between PC1 and PC2, were as follows: 58.5 % C_3S; 7.5 % C_2S; 7.5 % C_3A; and 11.5 % C_4AF; density 3.06 and BSS 325), distilled water and a water-cement ratio of 0.5. The paste was mixed manually for 2 minutes in a porcelain crucible and poured into the viscometer annulus.

The rheological measurements were taken with a Haake Roto Visco 1 rotational viscometer fitted with a Z38 DIN 53018 rotor spindle, a Z43 DIN 53018 graduated flask, a temperature control unit for coaxial cylindrical systems and a DC 30-B3 circulation thermostat [20].

The measurement schemes analysed, designed by combining angular velocity, stage duration and velocity step size (in rad/sec), are shown in Fig. 3.

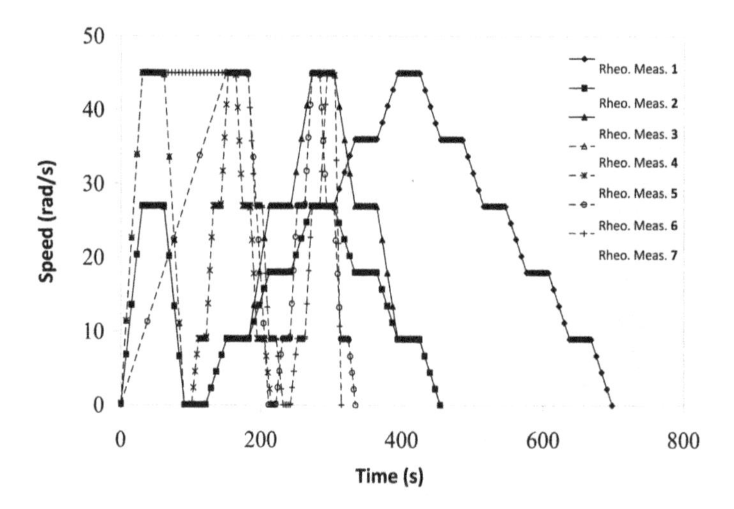

Figure 3. Rheological Measurement Schemes

Based on the results obtained in the preceding stage, a measurement scheme was designed and trials were conducted every 20 minutes (up to the time of the respective first nadir on the calorimetric curve [21-26] (Fig.3)) on two pure Portland cement pastes whose mineralogical composition was diametrically opposed. This compositional difference explained the difference in test time [21-26]: 120 minutes in paste PC1 and only 100 minutes in paste PC2, because 100 minutes is very near to 97 minutes 12 seconds ≈ 97 minutes, which was really the age of the 1st nadir for PC2. The physical-chemical characteristics of the two Portland cements are given in Table 1, along with their potential mineralogical composition, density and Blaine specific surface (BSS).

The two Portland cements exhibited similar fineness (Table 1 and Fig. 5), and the difference in their density could be partially attributed to their mineralogical composition. The water demand to prepare a standard consistency paste [36] was higher in PC1 than in PC2, a finding related to the capacity to form new hydrated compounds that required more chemically combined water. Setting times [36], in turn, were shorter in PC1 (Initial Setting Time: 200 minutes) than PC 2 pastes (IST: 270 minutes) [21-32]. This was also related to the formation of the new hydrated compounds that contribute to early age mechanical strength, primarily C_3A in this case [21-32][37]. The two Portland cements were also mixed with water at a w/c ratio of 0.5 and placed in warm (25ºC) water until tested.

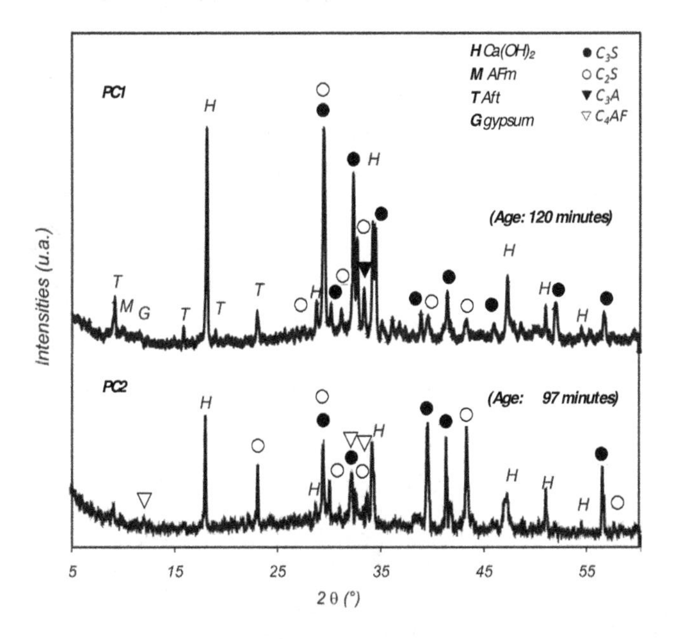

Figure 4. XRD analysis at the calorimetric curves nadir stages [16] for early hydration of *PC1* (1st nadir at 120 min. age) and *PC2* (1st nadir at 122 min. age)

In addition, the two PCs were analysed with the Frattini test [38] at very early ages: PC1 at 20, 40, 60, 80-100 and 120 minutes and PC2 at 20, 40, 60, 80 and 100 minutes (Figs. 6(a) and (b)). This test [38] is valid for pozzolanic or type CEM IV cements only [35]. Nonetheless, even though neither PC had to pass it, both were tested to compare their [CaO], [OH^-] and liquid phase pH at such early ages to better explain and understand their behaviour in the subsequent rheology tests. In the Frattini test for POZC, the calcium hydroxide content in the aqueous solution in contact with the hydrated sample kept at 40 ºC for 2, 7 and 28 days was compared to the solubility isotherm for calcium hydroxide in an alkaline solution kept at the same temperature. The mineral addition is regarded to produce pozzolanic activity (= positive result) when the calcium hydroxide concentration in the sample solution was below the solubility isotherm curve, but as both PC are plain, i.e., both of them had not any pozzolan amount and, for this reason, the [Ca] and [OH^-] contents of their respective liquid phase have to be, in contrast, necessarily over the solubility isotherm curve for calcium hydroxide in alkali solution (= negative result). The findings are given in Figs. 6(a) and (b).

Materials Parameters	Portland cements	
	PC1	PC2
C_3S, %	51	79
C_2S, %	16	2
C_3A, %	14	0
C_4AF, %	5	10
Álcalis eq., %	1.5	0.4
Density (Kg/l)	3.08	3.21
SSB, m^2/kg	319	301
Water demand, w/c	0.31	0.28
Initial setting point, min	3:20	4:30
Final setting point, min	5:10	6:15

Table 1. Chemical composition and physical parameters of two Portland cements

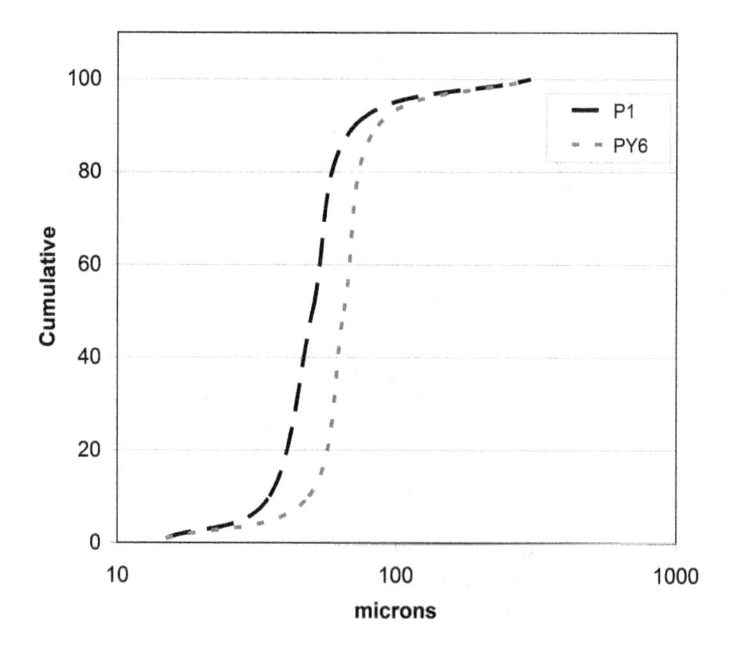

Figure 5. Particle size distribution of two Portland cements (Laser granulometry)

The effect of different mineralogical composition on the zeta potential of cement P1 and P2 was determined on a Malvern Instruments ZETASIZER 2000 particle sizer. The measuring principle consisted of forcing the particles to be studied across an electric field using laser light scattering techniques. After the bombardments of the particles, the refracted laser will be collected by a correlator in order to transform the data of measurement of the electric potential of double-layer of particle. An aqueous solution in contact with the hydrated sample were injected into the analyser. After stirring, the solutions introduced in the injection equipment to various ages: 20, 60, 120 minutes, initial and final setting time, first minimum and second maximum of the calorimetric curve. Then conducted a total of nine measurements for each solution.

3. Results and discussion

3.1. Frattini test: [CaO], [OH⁻] and liquid phase pH values (Figs. 6(a) and (b))

The Frattini test findings, (Fig. 6(b)), confirmed that both PCs were pure Portland type materials, for neither passed the test at any age (nor would they pass it at any age, naturally), as all the pastes were above the solubility isotherm for $Ca(OH)_2$ in an alkaline solution. The test nonetheless afforded the following information.

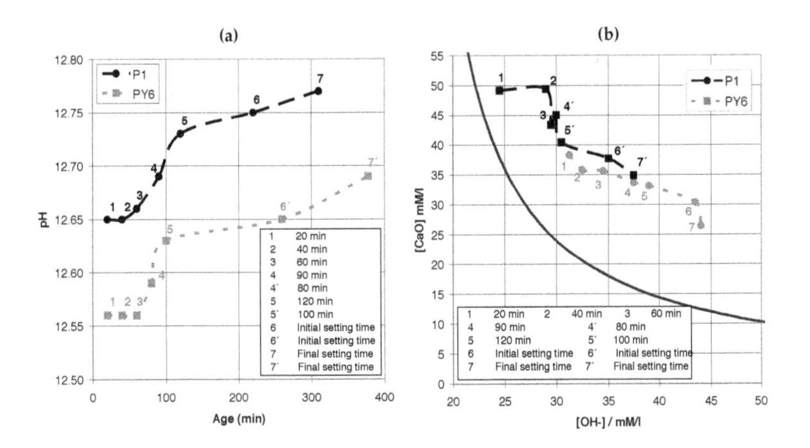

Figure 6. pH values and Frattini test results of the liquid phase for the two PC: PC1 and PC2

1. Regardless of the test age considered and all else being equal, the [CaO] values were consistently higher and the [OH⁻] values were consistently lower in PC2 than in PC1. The reason was the enormously different mineralogical composition of the two cements (Table 1), with PC2 having 79 % C_3S, 2 % C_2S, 0 % C_3A and 0.56 % $Na_2Oeq.$, and PC1 51 % C_3S, 16 % C2S, 14 % C_3A and 1.24 % Na_2Oeq. Consequently, more hexagonal crystal portlandite precipitated in the PC2 than the PC1 paste during hydration. By contrast, more KOH and NaOH were generated in PC1. Being much more soluble and exhibiting

greater exchange capacity [39], they remained in the liquid phase at increasing concentrations, especially NaOH, raising the alkalinity and the pH values of that phase (Figs. 5(a) and (b)) as hydration moved forward. These findings should have been reflected as well in paste behaviour in the rheology test: all else being equal, the shear stress values would necessarily be consistently higher in paste PC2 than in paste PC1. This expected behaviour was in fact observed in the present study (see Table 2 and the final paragraphs of item 3.2).

2. The decline in [CaO] in the liquid phase during the test, along with the rise in the [OH^-] values, was more or less sinusoidal in both cements (Figs. 6(a) and (b)). These findings might well be mirroring PC hydration and portlandite formation. Initial mixing with water until each liquid phase reached supersaturation would be reflected as the segments more or less parallel to the X axis. The concomitant precipitation of portlandite crystals in each paste would appear as the segments sloping more or less downward on the two curves. This would be followed by further hydration until the liquids again became supersaturated, and subsequent precipitation of more portlandite. This process would continue in both PCs until their entire stock of C_3S and C_2S was depleted.

Since neither NaOH nor KOH crystals would ever precipitate during hydration in this test [38], the respective pH values in the liquid phase of each paste would logically rise throughout hydration (Fig. 6(a)). And the higher the Na_2O (%) and K_2O (%) values in the PC, the steeper would that rise be. By contrast, the shear stress values in the respective paste or solid phase (Table 2) should have declined with rising alkali concentration. Both relationships, direct and indirect, were in fact found in this study (see Table 2 and the final paragraphs of item 3.2).

Time	Shear Stress (Pa) of PC1					Time	Shear Stress (Pa) of PC2				
(min.)	A	B	C	D	E	(min.)	A	B	C	D	E
20	62.35	32.61	6.41	31.86	7.54	20	163.1	76.10	14.55	50.06	12.86
40	71.09	36.98	6.03	34.78	8.16	40	168.0	81.95	14.61	47.94	13.41
60	79.33	48.03	11.01	44.52	9.97	60	176.2	77.65	13.78	47.76	12.93
80-100	93.12	56.12	14.09	46.57	13.18	80	172.5	79.11	12.39	42.55	11.38
120	91.91	54.07	12.30	46.15	11.35	100	162.5	75.78	12.04	42.87	12.83

Table 2. Most important Shear Stress values of two Portland cements

All the above findings were observed, moreover, despite the higher C_3A content in PC1 (14 %) than in PC2 (0 %). As a result,

- together with the formation and precipitation of portlandite crystals, prismatic, likewise hexagonal ettringite or *AFt* phase needles should also form in paste PC1; these would later evolve into hexagonal *AFm* phase needles (Fig. 4), in addition to CSH gels generated by C_3S and C_2S; and

- the initial setting time of PC1 should be shorter [36].

Despite these two unfavourable circumstances, however, its shear stress values never exceeded the values recorded in identical conditions in paste PC2, since by that time its liquid phase was much more basic than the PC2 liquid phase.

3.2. Rheological test findings

The flow curves for rheological measurements 1 to 7 made on CEM I [35] cement paste are shown in Fig. 7. In measurement 1, which had five angular velocities rather than the three in the other measurements, higher shear was required in the downward than in the upward arm, an indication of the formation of new bonds in the interim, leading to greater strength than at the beginning of the trial. A comparison of measurements 2 and 3 reveals the effect of the step size between stages: more bonds were broken when the step was larger. The initial stage was the same in measurements 1, 2 and 3: 30 seconds at 27 rad/sec.

The difference between measurements 3 and 4 lay in the initial velocity. An increase from 27 to 45 rad/sec was found to generate a shorter hysteresis cycle.

In measurements 1 to 4, the duration of each stage was 30 seconds, whereas in measurement 5 the duration was 10 seconds. Shortening the stage duration led to an overlap between the last ramp up stage and the first ramp down stage.

In all the aforementioned cases, one velocity was ramped to the next quickly. In measurement 6, however, the angular velocity of the initial stage was raised more slowly. Under these conditions, shear strength was found to be higher and the upward and downward arms to have a steeper slope, as a comparison of measurements 5 and 6 shows

Lastly, in measurement 7 the initial stage was lengthened to 150 seconds and the ramp down was begun immediately, followed by the ramp up. In this case greater stress was observed in the ramp up, which stood as proof that fewer bonds were broken during the ramp down that preceded it.

After analysing the effect of each of the aforementioned parameters, a measurement scheme was designed with 10-second stages, three velocities and down ramping proceeding up ramping. The scheme designed, shown in Fig. 8 with five singular points, was as follows: 10″ from 0 to 45 rad/sec; 30″ at 45; 10″ from 45 to 27; 10″ at 27; 10″ from 27 to 9; 10″ at 9; 10″ from 9 to 0; 10″ at rest; 10″ from 0 to 9; 10″ at 9; 10″ from 9 to 27; 10″ at 27; 10″from 27 to 45; 30″ at 45 and 10″ from 45 to 0. Therefore, singular points A, B, C, D and E, whose precise values for each PC tested are given in Table 2, were obtained from the measurement schemes shown in Fig. 7.

This measurement pattern was used to analyse the effect of Portland cement type (PC1 or PC2) and hydration time (at a w/c ratio of 0.5) on paste rheology. The rheological parameters were measured every 20 minutes until the first nadir appeared on the respective calorimetric curves, previously plotted with a conduction calorimeter at 25 ºC [21-26] and analysed by XRD technique as well (Fig. 4). The rheological results obtained in this study are given in Figs. 9 and 10.

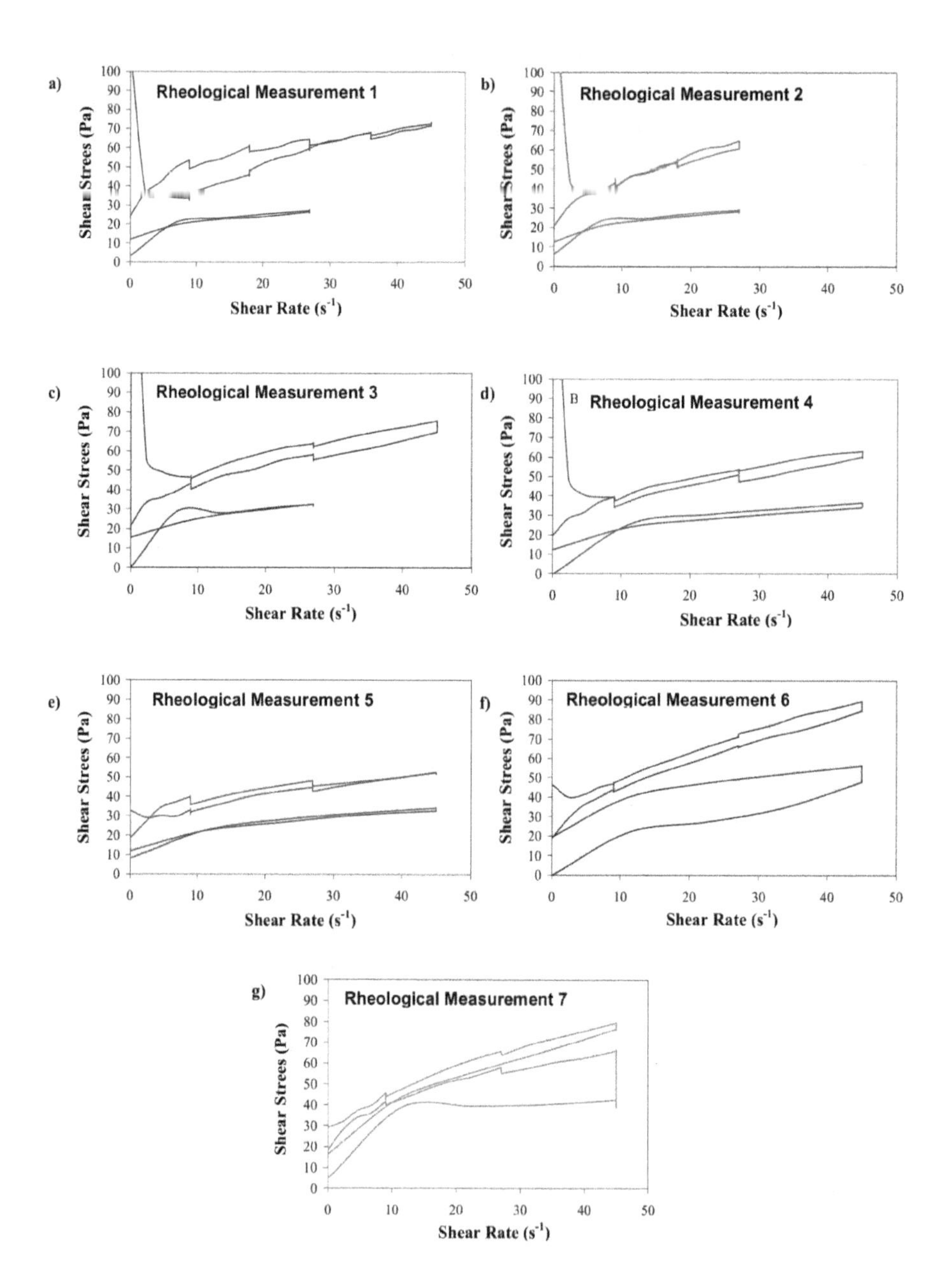

Figure 7. Flow curves of the seven rheological tests

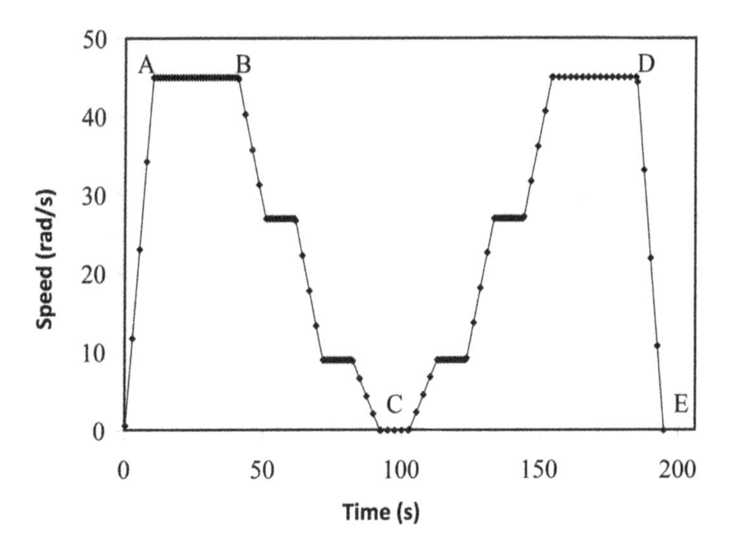

Figure 8. Rheological test finally selected for the two Portland cement: PC1 and PC2

Given the potential mineralogical composition of PC1 (Fig. 9) and due very likely to the formation and precipitation of ettringite hexagonal prismatic needles and rosettes and hexagonal portlandite and phase *AFm* platelets [27-32] (Fig. 4), the initial shear stress at point A rose from one measurement to the next as the hydration reactions progressed. Shear resistance declined substantially (up to 30-38 Pa) (thixotropic behaviour) at point B at all measurement times, since the angular velocity was held constant up to that point, although the general pattern was the same as for point A (increase from one measurement to the next as hydration progressed). Despite the removal of the mechanical force deriving from the spin transmitted by the rotor to the paste at point C, the paste nevertheless retained a certain amount of stress, which, as in point A, rose as hydration progressed. This was very likely due to the rotational inertia that remained in the paste particles (numerous and varied), even though the rotor had been stopped completely for 10" before reaching point C. At point D, then, the stimulus was the same as at point B, although the shear resistance exhibited by the paste was 1 to 10 Pa smaller at the former for the above reasons. This infers that most of the bonds, which initially had a stiffening effect (even if only due to gravity) because of the static position of the solid particles in the paste at the outset, and which must have been broken in B, were restored. This restoration must logically have occurred more speedily at earlier ages, given the lower force needed. Lastly, at point E, at the end of the test, remnant shear stress was recorded, as at point C. But in this case the values were greater at 20 and 40 minutes and smaller at 60, 80-100 and 120 minutes, a logical result, for the pastes were closer by then to their initial setting time (3h:20m) and ettringite, portlandite, phase *AFm* and CSH gel formation and precipitation should have increased.

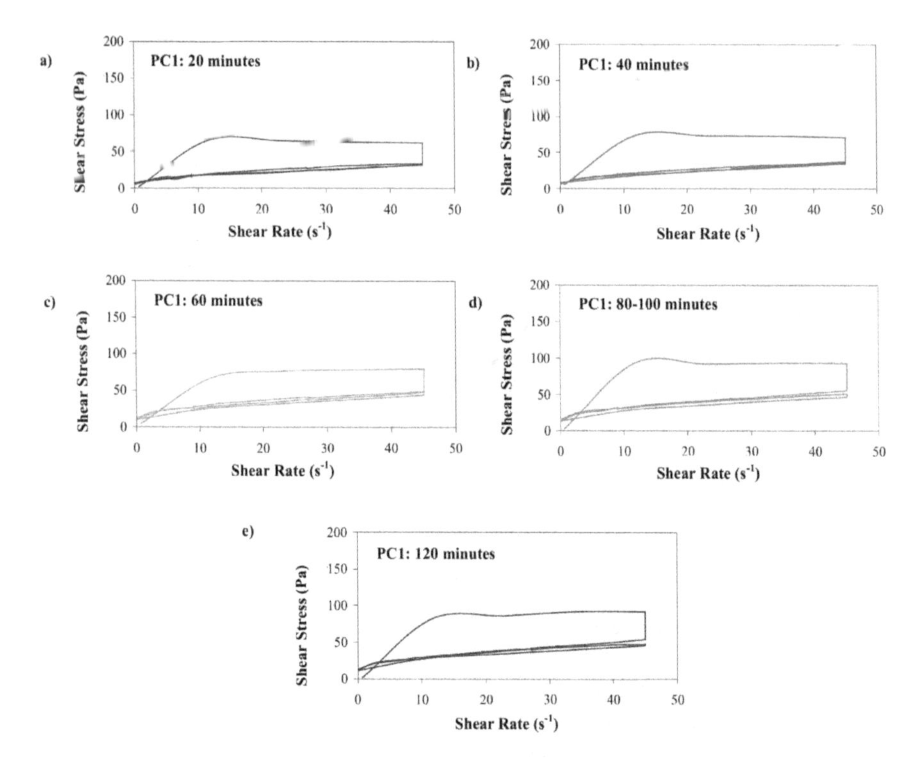

Figure 9. PC1 sequence of responses

The initial shear stress at point A likewise grew from one measurement to the next during the first hour of PC2 hydration (Fig. 10), after which it declined. Shear stress was found to be much greater than in PC1 due to the mineralogical composition, and consequently, to the nature of the many and diverse reaction products forming in the first few minutes. As in PC1, shear strength was observed to be lower (thixotropic behaviour), in this case by 87 to 100 Pa, at point B than at point A at all the test times and for the same reason, although the value climbed throughout the trial. In both PC1 and PC2, shear strength declined at point B by around one half of the value reached at point A. At point C, when the rotational force induced by the rotor and transmitted to the paste was completely removed, the PC2 paste was also observed to retain some stress, although with a clearly downward trend over time. As in cement PC1, at point D, the stimulus was the same as at point B, although the shear resistance exhibited by the paste was smaller (26 to 37 Pa). This would mean that some of the bonds which initially had a stiffening effect (even if only due to gravity) because of the static position of the solid particles in the paste at the outset, and which must have been broken in B, were restored, although this restoration was less intense than in PC1. Lastly, at point E, at the end of the trial, remnant shear stress was recorded, as at point C, but with lower values up to 80 minutes and higher values thereafter. This would have been expected because the setting time for this cement was longer, 4h:30m. Nonetheless, the behaviour was totally different from PC1 at this point, likewise as would have been expected in light of its totally different mineralogical composition.

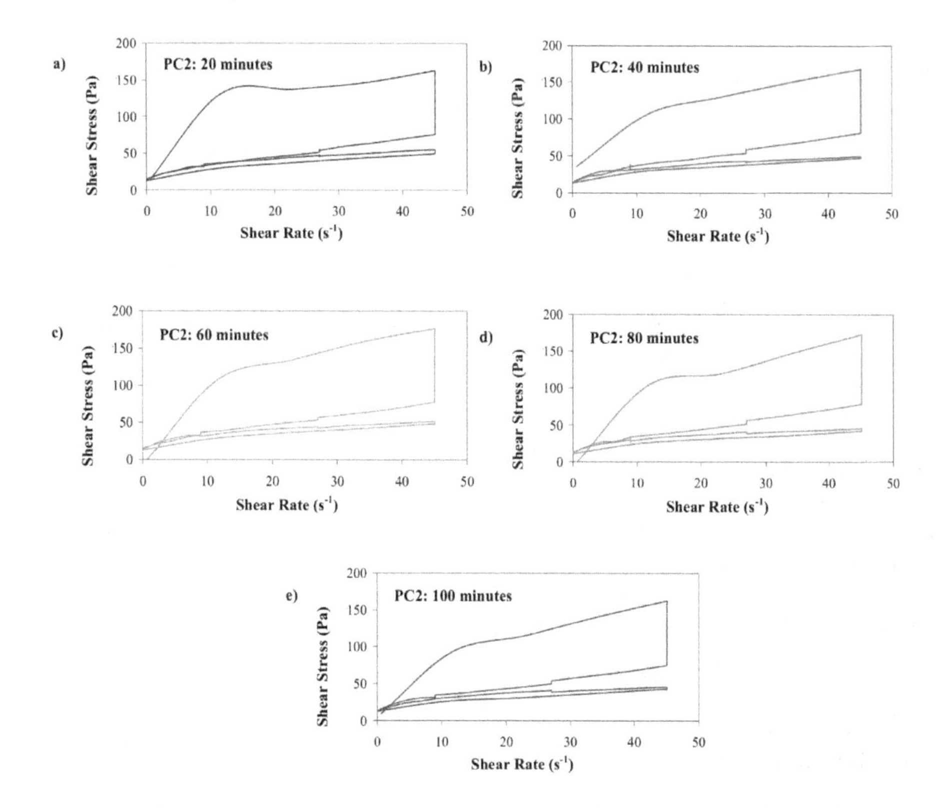

Figure 10. PC2 sequence of responses

Moreover, point by point at any given test age, the shear stress value in PC1 was consistently lower than in PC2. The reason would necessarily have been the hydroxy-induced [OH^-] alkalinity; consequently, the pH of the liquid phase in PC1 was always higher than in PC2.

Lastly, at all five points, the shear stress on PC1 rose as hydration moved forward, but only up to the age of 80-100 minutes, after which it declines. In the PC2 paste, by contrast,

- at point A stress grew up to the age of 60 minutes, to decline thereafter through the end of the trial,
- at point B stress followed a sinusoidal pattern, i.e., rising between 20 and 40 minutes and alternately rising and falling from that time on,

- at point C (1st remnant) stress grew up to the age of 40 minutes and then declined steadily through the end of the rheological trial,
- at points D and E (2nd remnant):
- in PC1 paste, as noted, stress rose up to the age of 80-100 minutes and then declined through the end of the trial, at 120 minutes, and
- in the PC2 paste, the stress values followed a sinusoidal pattern from the beginning to the end of the trial, rising in the final, 100-minute paste.

The reason for such a broad difference in the two cements' rheological behaviour, particularly as regards points D and E, between the last but one and the last age (80-100 to 120 minutes in PC1; 80 to 100 minutes in PC2), must have been that at those ages, the alkalinity values ([OH-] and pH) of their liquid phases were more widely separated than at any other. In other words, pH was lower than at any other age in paste PC2 and higher than at any other age in paste PC1 (see Fig. 6), even though the end of the rheological trial, 120 minutes, is nearer to its initial setting time (IST: 200 minutes) in PC1, than 100 minutes in PC2 (its IST is 270 minutes). All the foregoing was the result, in turn, of the differences in the potential chemical composition of the two PCs, here specifically in terms of their Na_2Oeq (%) content, which was 1.24 % > 0.6 % in PC1 (not a low alkali cement) [40] and 0.56 % < 0.6 % in PC2 (a low alkali cement). This, together with the higher C_3S content in the latter than in PC1 (79 compared to 51 %), would have induced more intense precipitation of microscopic hexagonal portlandite crystals and CSH gels in the solid phase of the paste, as a result of the permanent supersaturation in the liquid phase. That in turn must have raised the initial mass of the paste. Along with the hexagonal shape of the portlandite crystals, such increased mass would have led to a rise in the shear stress, reversing the decline recorded up until that time, contrary to the behaviour observed in PC1 in this regard.

Consequently, the PC1 paste behaved more uniformly at all points and ages than the PC2 paste.

3.3. Electrokinetic study

The zeta potential test findings for several pastes of Portland cements PC1 and PC2 are shown in Figure 11(a). Moreover, this figure also shows the results of conductivity (Fig. 11 (b)) and ionic mobility (Fig. 11 (c)) of all the pastes of both Portland cements.

In Figure 11 (a) shows the zeta potential at any age Portland cement PC1 was lower than for the case of Portland cement PC2. This behavior is related to the pH values obtained (Figure 6), since the more basic the pH of the sample is more negative potential corresponding value of Z (Fig. 12). Again, the reason was that regardless of the test age considered and all else being equal, the [CaO] values were consistently higher and the [OH^-] values were consistently lower in PC2 than in PC1, due to the different mineralogical composition of the two cements (Table 1). Consequently, more hexagonal crystal portlandite precipitated in the PC2 than the PC1 paste during hydration. By contrast, more KOH and NaOH were generated in PC1. Being much more soluble and exhibiting greater exchange capacity [39], they remained in the liquid phase at increasing concentrations, especially NaOH, raising the alkalinity and the pH values of that phase (Figs. 6(a) and (b)) as hydration moved forward.

The zeta potential is determined by the nature of the particle surface and the dispersion region, and as mentioned above pH is often an important parameter. Figure 12 shows an example of the variation of zeta potential and pH, The curve crosses the X axis, this point is called the isoelectric point. This means that the particles do not experience repulsion, so that the particles agglomerate. In fact there may be some attraction near this value as well, and as a rule if you want to ensure that there is repulsion between particles we must ensure that the value of zeta potential is greater than +30 mV and-30mV. Therefore, two Portland cements exhibit in the instability area of potential Z until the hydration reactions are completed and the pH is increased to stabilize. Therefore zeta potential values of the two Portland cements become increasingly approaching more negative to stable region.

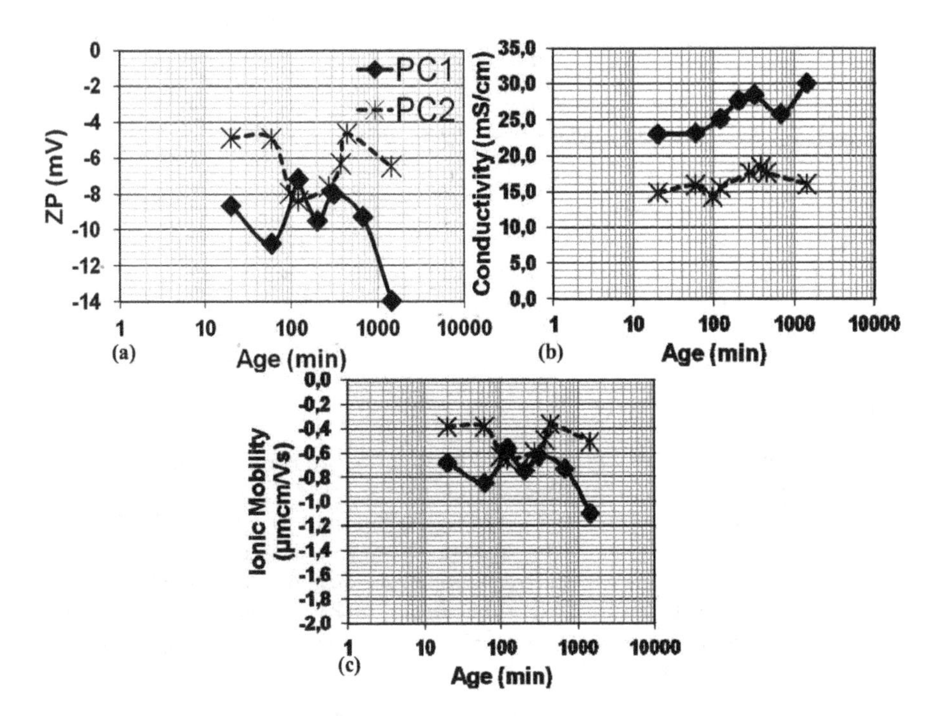

Figure 11. (a) Z Potential values of two Portland cements used PC1 and PC2. (b) Conductivity of two Portland cements used both PC1 and PC2. (c) Ionic Mobility of two Portland cements used both PC1 and PC2.

Therefore, PC1 is the Portland cement that reached the highest values of zeta potential causing further dispersion of individual particles. This means that the system approaches the stable area and consequently the corresponding viscosity of the pastes is greater PC2 cement. All the foregoing is the result, in turn, of the differences in the potential chemical composition of the two PCs, PC1 (not a low alkali cement) and PC2 (a low alkali cement). This statement is acuared with the water demand to prepare a standard consistency paste [36], was higher in PC1 than in PC2, a finding related to the capacity to form new hydrated compounds that required more chemically combined water. Setting times [36]. The higher C_3S content in the latter than in PC1, would have induced more intense precipitation of microscopic hexagonal portlandite crystals and CSH gels in the solid phase of the paste, as a result of the permanent supersaturation in the liquid phase. So would have led to a rise in the shear stress, reversing the decline recorded up until that time, contrary to the behaviour observed in PC1 in this regard.

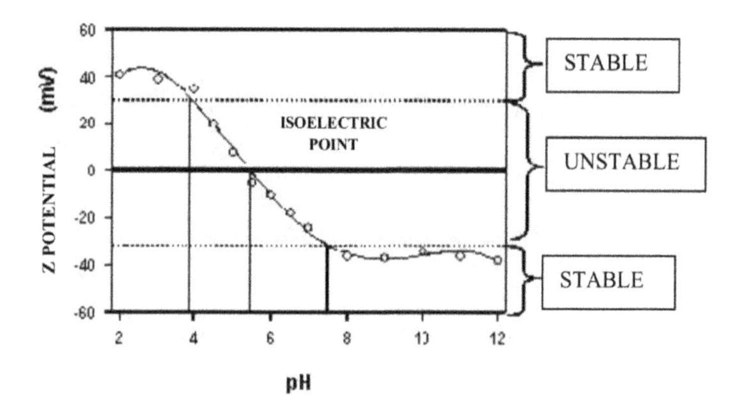

Figure 12. Dependence of Zeta Potential with pH

In Figure 11 (b) shows the evolution of the conductivity of two Portland cements with age. The differences in the two cements are mainly due to their very different mineralogical composition, specifically in terms of their Na_2Oeq (%) content, which was 1.24 % > 0.6 % in PC1 (not a low alkali cement) [40] and 0.56 % < 0.6 % in PC2 (a low alkali cement). Moreover, the graphs of the two Portland cements having the same pattern, however, the Portland cement PC2 lags the cement PC1, because the rate of hydration of portland cement mineral components PC1 is increased. The increased speed of hydration corresponding to C_3A, and the PC1 cement contains about 14%.

Furthermore, in connection with the ionic mobility (Figure 11 (c)) is met and the above explained through the graph shown in Figure 12. The ions present in the Portland cement samples from PC2 have higher ionic mobility. This trend is related to the zeta potential results obtained for the two Portland cements as for Portland cement PC1 values are closer to the stable zone where the ions are more dispersed. By contrast, the values for the Z potential of Portland cement PC2 are closer to the unstable zone, so that different ions tend to agglomerate, thus its ion mobility is higher.

4. Conclusions

The following conclusions may be drawn from this study.

1. Rheological measurement design has a substantial impact on the type of response. Consequently, a given stimulus may yield different shear strength values as well as both *thixotropic* and *anti-thixotropic* behaviour, depending on the measurement scheme used.

2. The following findings were observed in connection with the mineralogical composition of Portland cement and its effect on the nature and development of the many and diverse hydration products forming in the initial phases of hydration of the fresh paste.

a. Similar behaviour: ≈ 50% decline in shear stress after 30″ of rotation and further loss when the stimulus was repeated

b. Clearly different behaviour: higher shear strength in paste PC2, greater bond recovery in paste PC1

c. Remnant shear stress (points C and E) in paste PC1 that declined throughout the trial; in paste PC2, the values at point C also declined, whereas at point E they rose, as would be expected, further to conclusion 3.

3. The explanation for the preceding conclusions, in particular conclusion 2, lies in the differences in [CaO] and $[OH^-]$ in the liquid phases of the two PCs and their concomitant pH values. These, in turn, were the result of the differences in their mineralogical composition, especially with respect to their N_2O (%) and K_2O (%) contents, which determined the higher alkalinity in PC1 and the lower shear strength of the paste made from that material.

4. PC1 is the cement that reached the highest values of zeta potential causing further dispersion of individual particles. This means that the system approaches the stable area and consequently the corresponding viscosity of the pastes is greater PC2 cement. All the foregoing is the result, in turn, of the differences in the potential chemical composition of the two PCs.

Author details

R. Talero[1*], C. Pedrajas[1] and V. Rahhal[2]

*Address all correspondence to: rtalero@ietcc.csic.es

1 Eduardo Torroja" Institute for Construction Sciences – CSIC; Calle Serrano Galvache; Madrid, Spain

2 Departamento de Ingeniería Civil Facultad de Ingeniería UNCPBA. Av. del Valle, Argentina, Olavarría, Argentina

References

[1] Uchikawa U. Influence of pozzolan on hydration of C_3A 7th International Congress on the Chemistry of Cement, Proceedings, Vol. IV, pp 24/29, 1980, Paris – France.

[2] Uchikawa H, Uchida S, Hanehara S. Effect of character of glass phase in blending components on their reactivity in calcium hydroxide mixture. 8th International Congress on the Chemistry of Cement, Proceedings, Vol – III, pp 245/250, 1986, Río de Janeiro – Brasil.

[3] Husson S, Gullhot B, Pera J. Influence of different fillers on the hydration of C_3S. 9th International Congress on the Chemistry of Cement, Proceedings, Vol. IV, pp 83/89, 1992. New Delhi – India.

[4] Bombled J.P. Etude rhéologique des pâtes crues de cimenterie. Revue des Matériaux de Contruction 1970;(609):229-38.

[5] Miranda J, Flores-Alés V, Barrios J. Aportaciones al estudio reológico de pastas y morteros de cemento Portland. Mater. Construc. 2000;50(257):47-55.

[6] Martínez de la Cuesta P.J., Rus Martínez E, Díaz Molina F, Luna Blanco S. Reología de mezclas de cemento con fíller dolomítico. Mater. Construc. 2000;50(258):11-25.

[7] Puertas F, Alonso M.M., Vázquez T. Efecto de un aditivo basado en policarboxilatos sobre la reología y el fraguado de pastas de cemento Pórtland. Cemento y Hormigón 2002;(844):4-12.

[8] Bombled J.P. Rhéologie des mortiers et des bétons frais. Etudes de la pâte interstitielle de ciment. Revue des Matériaux de Contruction 1974;(688):137-55.

[9] Caufin B, Papo A. Rheological behaviour of cement pastes. Zement-Kalk-Gips. 1984;12:656-61.

[10] Tattersall, G.H., Banfill, P.F.G. The rheology of fresh concrete. Pitman, 1983, pp 356.

[11] Barnes, H.A. A handbook of elementary rheology, Institute of Non–Newtonian Fluid Mechanics, University of Wales, (2000), 200pp.

[12] Barnes, H.A., Hutton, J.F., Walters, K. An introduction to rheology, Elsevier, (1989), 199pp. Yahia, A., Khayat, K.H. Analytical models for estimating yield stress of high performance pseudoplastic grout, Cement and Concrete Research, vol.31, (2001), pp. 731–738.

[13] Yahia, A., Khayat, K.H. Analytical models for estimating yield stress of high performance pseudoplastic grout, Cement and Concrete Research, vol.31, (2001), pp. 731–738.

[14] Banfill, P.F.G., Saunders, D.C. On the viscometric examination of cement pastes, Cement and Concrete Research, vol.11, (1981), pp.363–370.

[15] Hattori, K. Izumi, K. A rheological expression of coagulation rate theory, Parts 1–3, Journal of Dispersion Science and Technology, vol.3, (1982), pp.129–145, pp.147–167, pp.169–193.

[16] Tattersall, G.H The rheology of portland cement pastes, British Journal of Applied Physics, vol.6, (1955), pp.165–167.

[17] Banfill, P.F.G. A viscometric study of cement pastes including a note on experimental techniques, Magazine of Concrete Research, vol.33, (1981), pp.37–47.

[18] Orban, J., Parcevaux ,P., Guillot, D. Influence of shear history on the rheological properties of oil well cement slurries, 8th International Congress on the Chemistry of Cement, vol. 6, (1986), pp.243–247.

[19] Hodne, H., Saasen, A., O'Hagan, A.B., Wick, S.O. Effects of time and shear energy on the rheological behaviour of oilwell cement slurries, Cement and Concrete Research, vol.30, (2000), pp.1759–1766.

[20] Criado Sanz M, Palomo Sánchez A, Fernández Jiménez A. Nuevos materiales cementantes basados en cenizas volantes. Influencia de los aditivos en las propiedades reológicas. Monografía 413 Instituto de Ciencias de la Construcción Eduardo Torroja, Madrid, España, 2006.

[21] Rahhal V, Cabrera O, Talero R. Calorimetry of portland cement with silica fume and gypsum additions. J. Therm. Anal. Cal. 2007;87(2):331-36.

[22] Talero R, Rahhal V. Influence of *aluminic* pozzolans, quartz and gypsum additions on Portland cement hydration.- 12th International Congress on the Chemistry of Cement. Proceedings, Montreal–Canada, 8–13 july 2007.

[23] Rahhal V, Talero R. Calorimetry of portland cement with metakaolins, quartz and gypsum addtions J Therm Anal Cal 2008;91(3):825-34.

[24] Talero R, Rahhal V. Calorimetric comparison of portland cement containing silica fume and metakaolin: Is silica fume, like metakaolin, characterized by pozzolanic activity that is more *specific* than *generic*? J Therm Anal Cal 2009;2:383-93.

[25] Rahhal V. Talero R. Calorimetry of Portland cement with silica fume, diatomite and quartz additions. Construction and Building Materials 2009;23:3367-74.

[26] Rahhal V, Bonavetti, V, Trusilewicz L, Pedrajas C, Talero R. Role of the filler on portland cement hydration at early ages. Construction and Building Materials. Ref. No.: CONBUILDMAT-D-11-00171 (corrected and sent for being accepted for publication).

[27] Talero R. Expansive Synergic Effect of ettringite from pozzolan (metakaolin) and from OPC, co-precipitating in a common plaster-bearings solution. Part I: By cement pastes and mortars. Construction and Building Materials 2010;24:1779-89.

[28] Talero R. Expansive Synergic Effect of ettringite from pozzolan (metakaolin) and from OPC, co-precipitating in a common plaster-bearings solution. Part II: Funda-

mentals, explanation and justification. Construction and Building Materials 2011;25:1139-58.

[29] Talero R. Kinetochemical and morphological differentiation of ettringites by the Le Chatelier-Ansttet test. Cem Concr Res. 2002;32:707–17.

[30] Talero R. Performance of metakaolin and portland cements in ettringite formation as determined by ASTM C 452-68: kinetic and morphological differences. Cem Concr Res 2005;32:1269–84.

[31] Talero R. Kinetic and morphological differentiation of ettringites by metakaolín, Portland cements and ASTM C 452-68 test. Part I: Kinetic differentiation. Mater Constr 2008;58 (292):45-68.

[32] Talero R. Kinetic and morphological differentiation of ettringites by metakaolin, portland cements and ASTM C 452-68 test. Part II: Morphological differentiation by SEM and XRD analysis. Mater Construí. 2009;59(293):17-34.

[33] Talero R. Is the clay *exchange capacity* concept wholly applicable to pozzolans? Mater. Constr. 2004;54(276);17-36, and 2005;55(277):82.

[34] J.P. Bombled: Etude rhéologique des pâtes crues de cimenterie.- Revue de Matériaux de Construction 1970;609:229-38.

[35] Instrucción para la Recepción de Cementos RC-08 (R.D.956/2008, de 6 de junio; BOE núm. 148 del 16 de junio de 2008).

[36] EN 196-3 Standard. Methods of testing cement. Part 3. Times of setting and volume stability determinations. AENOR, Calle Génova, 6, 28004 – MADRID – Spain.

[37] EN 196-1 Standard: Methods of testing cements Part 1. Mechanical Strengths determination. AENOR.

[38] EN 196-5 Standard: Methods of testing cements; Part 5. Pozzolanicity test for POZC. AENOR, Calle Génova No. 6; 28004-MADRID-Spain ≈ Pliego de Prescripciones Técnicas Generales para la Recepción de Cementos RC-75 (Decreto de la Presidencia del Gobierno 1964/1975 de 23 de mayo– B.O:E. nº 206 de 28 de agosto de 1975) = Frattini, N. Solubilità dell'idrato di calcio in presenza di idrato di potassio e idrato di sodio. *Annali di Chimica Applicata* 1949;39:616-20.

[39] Talero R. Is the clay "exchange capacity" concept wholly applicable to pozzolans? Mater. Construc. 2004;54(276):17-36; 2005;55(277):82.

[40] ASTM C 150-95 Standard: Standard Specification for Portland Cement.- ANNUAL BOOK OF ASTM STANDARDS, Section 4 Construction, Vol. 04.01, pp. 128-132, 1995.

Chapter 5

Unsteady Axial Viscoelastic Pipe Flows of an Oldroyd B Fluid

A. Abu-El Hassan and E. M. El-Maghawry

Additional information is available at the end of the chapter

1. Introduction

The unsteady flow of a fluid in cylindrical pipes of uniform circular cross-section has applications in medicine, chemical and petroleum industries [3,4,5]. For viscoelastic fluids, the unsteady axial decay problem for UCM fluid is considered by Rahman et al. [6]; and for Newtonian fluids as a special case. Rajagopal [7] has studied exact solutions for a class of unsteady unidirectional flows of a second-order fluid under four different flow situations. Atalik et al. [8] furnished a strong numerical evidence that non-linear Poiseuille flow is unstable for UCM, Oldroyd-B and Giesekus models. This fact is supported experimentally by Yesilata, [9]. The unsteady flow of a blood, considered as Oldroyd-B fluid, in tubes of rigid walls under specific APGs is concerned by Pontrelli, [10, 11].

Flow of a polymer solution in a circular tube under a pulsatile APG was investigated by Barnes et al. [12, 13].The same problem for a White-Metzner fluid is performed by Davies et al. [14] and Phan-Thien [15]. Recently, periodic APG for a second-order fluid has been studied by Hayat et al. [16]. Numerical simulation based on the role of the pulsatile wall shear stress in blood flow, is investigated by Grigioni et al. [1].

The present paper is concerned with the unsteady flow of a viscoelastic Oldroyd-B fluid along the axis of an infinite tube of circular cross-section. The driving force is assumed to be a time-dependent APG in the following three cases:

i. APG varies exponentially with time,

ii. Pulsating APG,

iii. A starting flow under a constant APG.

2. Formulation of the problem

The momentum and continuity equations for an incompressible and homogenous fluid are given by

$$\rho \frac{d\underline{q}}{dt} = -\nabla P + \nabla \cdot \underline{\underline{S}}, \tag{1}$$

and

$$\nabla \cdot \underline{q} = 0, \tag{2}$$

where ϱ is the material density, $\underline{q}$ is the velocity field, p is the isotropic pressure and $\underline{S}$ is the Cauchy or extra-stress tensor. The constitutive equation of Oldroyd-B fluid is written as

$$\underline{\underline{T}} = -p\underline{\underline{I}} + S; \underline{\underline{S}} + \lambda_1 \overset{\nabla}{\underline{\underline{S}}} = \mu\{\underline{\underline{A_1}} + \lambda_2 \overset{\nabla}{\underline{\underline{A_1}}}\} \tag{3}$$

where $\underline{T}$ is the total stress, $\underline{I}$ is the unit tensor, μ is a constant viscosity, λ_1 and λ_2, $(0 \le \lambda_2 \le \lambda_1)$ are the material time constants, termed as relaxation and retardation times; respectively. The deformation tensor $\underline{A}_1$ is defined by

$$\underline{\underline{A}}_1 = \underline{\underline{L}} + \underline{\underline{L}}^T; \underline{\underline{L}} = \nabla \underline{q}. \tag{4}$$

and "∇" denotes the upper convected derivative ; i.e. for a symmetric tensor $\underline{G}$ we get,

$$\overset{\nabla}{\underline{\underline{G}}} = \frac{\partial \underline{\underline{G}}}{\partial t} + \underline{q} \cdot \nabla \underline{\underline{G}} - \underline{\underline{G}} \cdot \underline{\underline{L}} - \underline{\underline{L}}^T \cdot \underline{\underline{G}}. \tag{5}$$

The symmetry of the problem implies that $\underline{S}$ and $\underline{q}$ depend only on the radial coordinate r in the cylindrical polar coordinates (r,θ,z) where the z-axis is chosen to coincide with the axis of the cylinder. Moreover, the velocity field is assumed to have only a z-component, i.e.

$$\underline{q} = (0, 0, \underline{w}), \tag{6}$$

which satisfies the continuity equation (2) identically. The substitution of Eq. (6), into Eqs. (1) and (3) yields the set of equations

$$S_{rz} + \lambda_1 \frac{\partial S_{rz}}{\partial t} = \mu\left(\frac{\partial w}{\partial r} + \lambda_2 \frac{\partial^2 w}{\partial r \partial t}\right), \tag{7}$$

$$\frac{\partial p}{\partial z} = \frac{\partial S_{rz}}{\partial r} + \frac{1}{r} S_{rz} - \rho \frac{\partial w}{\partial t}, \tag{8}$$

$$\frac{\partial p}{\partial r} = \frac{\partial p}{\partial \theta} = 0. \tag{9}$$

Equations (8) and (9) imply that the pressure function takes the form; $p = z\ f(t) + c$, so that

$$\frac{\partial p}{\partial z} = f(t). \tag{10}$$

The elimination of S_{rz} from (7) and (8) shows that velocity field $w(r,\ t)$ is governed by:

$$\rho\left(\lambda_1 \frac{\partial^2 w}{\partial t^2} + \frac{\partial w}{\partial t}\right) - \mu\left(1 + \lambda_2 \frac{\partial}{\partial t}\right)\left(\frac{\partial^2 w}{\partial t^2} + \frac{1}{r}\frac{\partial w}{\partial r}\right) = -\left(1 + \lambda_1 \frac{\partial}{\partial t}\right)\frac{\partial p}{\partial z}. \tag{11}$$

The non-slip condition on the wall and the finiteness of w on the axis give

$$w(r,t)|_{r=R} = 0 \text{ and } \frac{\partial w}{\partial r}|_{r=0} = 0. \tag{12}$$

Introducing the dimensionless quantities

$$\eta = \frac{r}{R}, \tau = \frac{\mu t}{\rho R^2}, \varphi = \frac{\mu L}{\Delta P R^2} w, \lambda = \frac{\lambda_2}{\lambda_1} \text{ and } H = \frac{\lambda_1 \mu}{\rho R^2} = \frac{We}{\text{Re}}, \tag{13}$$

where R is the radius of the pipe, ΔP a characteristic pressure difference, L is a characteristic length, We and Re are the Weissenberg and Reynolds numbers; respectively, into Eqs. (10), (11) and (12) we get

$$H\frac{\partial^2 \varphi}{\partial \tau^2} + \frac{\partial \varphi}{\partial \tau} - [1 + \lambda H \frac{\partial}{\partial \tau}][\frac{\partial^2 \varphi}{\partial \eta^2} + \frac{1}{\eta}\frac{\partial \varphi}{\partial \eta}] = [1 + H\frac{\partial}{\partial \tau}]\Psi(\tau), \tag{14}$$

with the BCs.

$$\varphi(1,\tau)=0 \text{ and } \frac{\partial\varphi(0,\tau)}{\partial\eta}=0, \tag{15}$$

and

$$\Psi(\tau)=-\frac{L}{\Delta p}\frac{\partial p}{\partial z}=-\frac{L}{\Delta p}f(t). \tag{16}$$

Equation (14) subject to BCs. (15) is to be solved for different types of APGs; i.e. different forms of the function $\Psi(\tau)$.

3. Pressure gradient varying exponentially with time

We consider the two cases of exponentially increasing and decreasing with time APGs separately.

3.1. Pressure gradient increasing exponentially with time

Let,

$$\Psi(\tau)=-\frac{L}{\Delta p}\frac{\partial p}{\partial z}=Ke^{\alpha^2\tau}, \tag{17}$$

and assume that

$$\varphi(\eta,\tau)=g(\eta)e^{\alpha^2\tau}, \tag{18}$$

where K and α are constants. The substitution of Eqs. (17) and (18) into Eq. (14) leads to

$$g''+\frac{1}{\eta}g'-\frac{\alpha^2(H\alpha^2+1)}{\lambda H\alpha^2+1}g=-K\frac{H\alpha^2+1}{\lambda H\alpha^2+1}, \tag{19}$$

while the BCs. (15) reduce to

$$g(1) = 0, g'(0) = 0 \tag{20}$$

A solution of Eq. (19) subject to the BCs. (20) is

$$g(\eta) = \frac{K}{\alpha^2}[1 - \frac{I_0(\beta\eta)}{I_0(\beta)}], \tag{21}$$

where I_0 (x) is the modified Bessel-functions of zero-order, and

$$\beta^2 = \frac{\alpha^2(1+H\alpha^2)}{1+\lambda H\alpha^2} \ . \tag{22}$$

Therefore, the velocity field is given by

$$\varphi(\eta,\tau) = \frac{K}{\alpha^2}[1 - \frac{I_0(\beta\eta)}{I_0(\beta)}]e^{\alpha^2\tau}. \tag{23}$$

The solution given by Eq. (23) processes the following properties:

i. The time dependence is exponentially increasing such that for $\eta \neq 1$ $\lim_{\tau\to\infty}\phi(\eta, \tau) \to \infty$. It may be recommendable to choose another APG which increases up to a certain finite limit in order to keep $\phi(\eta, \tau)$ finite.

ii. The present solution depends on the parameter β in the same form as the solution for the UCM [6]. For any value of β the Oldroyd-B fluid exhibits the same form as the UCM- fluid. However, in the present case β depends on λ in addition to H and α^2. A close inspection show that $\lim_{\lambda\to 0}\beta^2 = \beta^2$ for the UCM-fluid while the $\lim_{\lambda\to 1}\beta^2 = \alpha^2$ which coincides with the case of the Newtonian fluid, [8].

iii. The parameter β is inversely proportional to λ where the decay rate increases by increasing the value of H. However, as mentioned above, as λ approaches the value λ = 1 all the curves matches together approaching the value $\beta^2 = \alpha^2$ asymptotically. The behavior of β as a function of λ, where H is taken as a parameter is shown in Fig. (1).

For small values of $|\beta|$ and by using the asymptotic expansion of I_0 (x),

it can be shown that the velocity profiles approaches the parabolic distribution;

$$\lim_{\beta\to 0}\varphi(\eta,\tau) = \frac{K(H\alpha^2+1)}{4(\lambda H\alpha^2+1)}(1-\eta^2)e^{\alpha^2\tau} \tag{24}$$

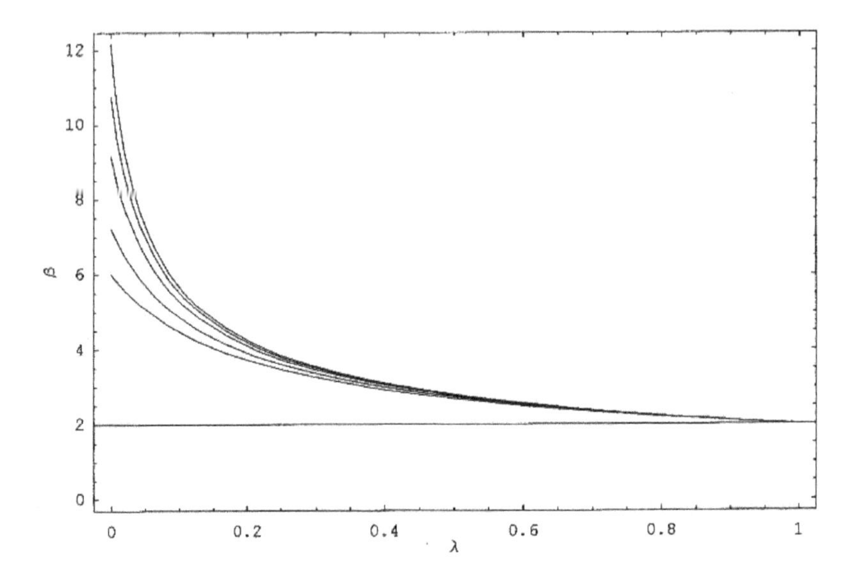

Figure 1. The $\lambda\beta$ - relation H= 2, 3, 5, 7, 9, (Bottom to top)

For the case of large $|\beta|$ the velocity distribution is given as;

$$\lim_{\beta\to\infty} \varphi(\eta,\tau) = \frac{K}{\alpha^2}[1 - \frac{1}{\sqrt{\eta}} e^{-\beta(1-\eta)}] e^{\alpha^2 \tau} \tag{25}$$

This solution is completely different from the parabolic distribution and it depends on η only in the neighborhood of the wall. Therefore, such a fluid exhibits boundary effects.

The rising-APG velocity field $\phi(\eta, \tau)$ is plotted in Figs. (2a) and (2b) as a function of η at different values of β for $\alpha = 2$ and $\alpha = 5$.

3.2. Pressure gradient decreasing exponentially with time

The solution at present is obtained from the previous case by changing α^2 by $-\alpha^{-2}$. Therefore,

$$\varphi(\eta,\tau) = -\frac{K}{\alpha^2}[1 - \frac{j_0(\beta_1\eta)}{j_0(\beta_1)}] e^{-\alpha^2\tau}. \tag{26}$$

where

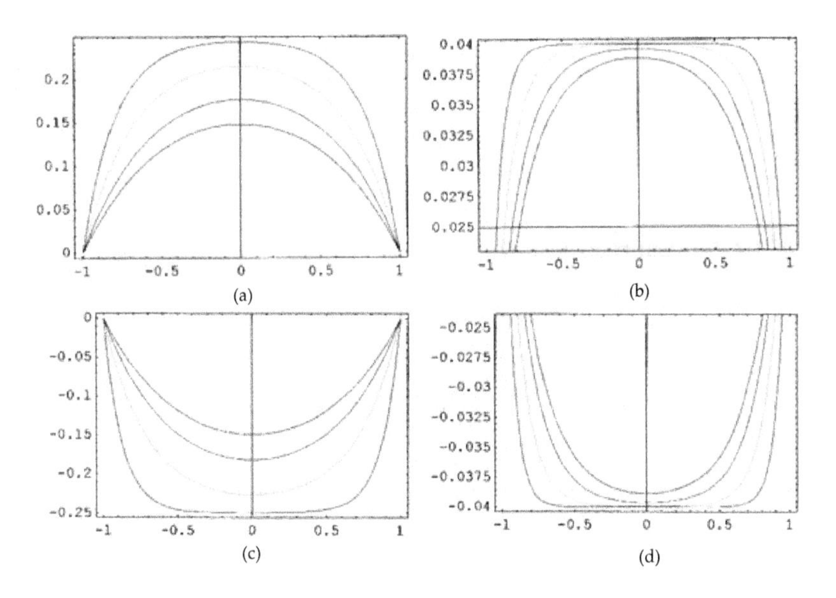

Figure 2. Rising – (a,b) APG velocity filed ;β=5.2, 3.5, 2.5, 2.1 (Bottom to top) Fig. (c) : Decreasing – APG velocity filed ;β=8.7, 3.9, 2.6, 2.1 (Top to Bottom) Fig. (d) : Decreasing – APG velocity filed ;β=16.4, 9.2, 6.5, 5.3 (Top to Bottom)

$$\beta_1^2 = \frac{\alpha^2(1-H\alpha^2)}{1-\lambda H\alpha^2} \tag{27}$$

The discussion of this solution is similar to the case of increasing APG except that the velocity decays exponentially with time and the value $\alpha^2 = 1/\lambda H$ is not permissible as it leads to infinite β_1^2 ; i.e.

$$\lim_{\alpha_1^2 \to 1/\lambda H} \beta_1^2 \to \infty \tag{28}$$

The two cases of small and large $|\beta_1|$ produce similar results as the previous solution. Thus

$$\lim_{\beta_1 \to 0} \phi(\eta,\tau) = -\frac{K}{4\alpha^2}\beta_1^2(1-\eta^2)e^{-\alpha\tau}, \tag{29}$$

and

$$\lim_{\beta_1\to\infty} \varphi(\eta,\tau) = -\frac{K}{\alpha^2}[1-\frac{1}{\sqrt{\eta}}\frac{\cos(\beta_1\eta-\frac{\pi}{4})}{\cos(\beta_1-\frac{\pi}{4})}]e^{-\alpha^2\tau}. \tag{30}$$

4. Pulsating pressure gradient

The present case requires the solution of Eq. (14) subject to BCs. (15) in the form

$$\Psi(\tau) = -\frac{L}{\Delta P}\frac{\partial p}{\partial z} = Ke^{in\tau}; \quad i=\sqrt{-1}, \tag{31}$$

K and n are constants. Assuming the velocity function has the form

$$\varphi(\eta,\tau) = re\left[f(\eta)e^{in\tau}\right], \tag{32}$$

$$\therefore f'' + \frac{1}{\eta}f' - in\frac{(1+inH)}{(1+in\lambda H)}f = -K\frac{(1+inH)}{(1+in\lambda H)}. \tag{33}$$

The solution of this equation satisfying the BCs. (15) is :

$$f(\eta) = \frac{k}{in}[1-\frac{I_0(\beta\eta)}{I_0(\beta)}], \qquad \beta^2 = in\frac{(1+inH)}{(1+in\lambda H)}. \tag{34}$$

Hence, the velocity distribution is given by:

$$\phi(\eta,\tau) = re\left\{\frac{k}{in}e^{in\tau}[1-\frac{I_0(\beta\eta)}{I_0(\beta)}]\right\}. \tag{35}$$

Obviously; for small $|\beta|$,

$$\lim_{\beta\to 0} f(\eta) = \frac{K}{in}\left(\frac{\beta^2(1-\eta^2)}{4}\right). \tag{36}$$

and for large $|\beta|$

$$\lim_{\beta\to\infty} \frac{I_0(\beta\eta)}{I_0(\beta)} = \frac{1}{\sqrt{\eta}} e^{-\beta(1-\eta)}, \tag{37}$$

So that,

$$\varphi(\eta,\tau) = re\left\{\frac{K}{in}[1 - \frac{1}{\sqrt{\eta}} e^{-\beta(1-\eta)} e^{in\tau}]\right\}, \tag{38}$$

where,

$$\beta^2 = \frac{in(1+inH)}{(1+in\lambda H)} = \frac{1}{1+n^2\lambda^2H^2}[n^2H(\lambda-1\} + in(1+n^2\lambda H^2)], \tag{39}$$

or simply,

$$\beta = \sqrt{\Re}\, e^{i\theta/2}, \tag{40}$$

$$\Re = \frac{n}{1+n^2\lambda^2 H^2}\sqrt{n^2H^2(1-\lambda)^2 + (1+n^2\lambda H^2)^2}, \tag{41}$$

$$\frac{\theta}{2} = -\frac{1}{2}\mathrm{Tan}^{-1}\left[\frac{1+n^2\lambda H}{nH(1-\lambda)}\right]. \tag{42}$$

Substituting from Eqs.(40,41,42) into Eq. (37), we get:

$$\lim_{\beta\to\infty}\varphi(\eta,\tau) = \frac{k}{n}\left\{\sin n\tau - \frac{1}{\sqrt{\eta}} e^{-(1-\eta)\sqrt{\Re}\cos(\theta/2)} \sin[n\tau - (1-\eta\sqrt{\Re}\sin\frac{\theta}{2})]\right\}. \tag{43}$$

As $\lambda \to 0$, [6], $\Re \to r_1 = n\sqrt{1+n^2H^2}$ and $\frac{\theta}{2} \to \frac{\theta_1}{2} = -\frac{1}{2}Tan^{-1}(\frac{1}{nH})$.

Then

$$\varphi(\eta,\tau) = \frac{k}{n}\left\{\sin n\tau - \frac{1}{\sqrt{\eta}} e^{-(1-\eta)\sqrt{r_1}(\cos\theta_1/2)} \sin[n\tau - (1-\eta)\sqrt{r_1}(\sin\frac{\theta_1}{2})]\right\} \tag{44}$$

The velocity field $\varphi(\eta,\tau)$ is plotted in Figs. (3a) and (3b); respectively, against η for different values of β. The two limiting cases for small and large $|\beta|$ are represented in three-dimensional Figs. (4a) and (4b) in order to emphasize the oscillating properties of the solution.

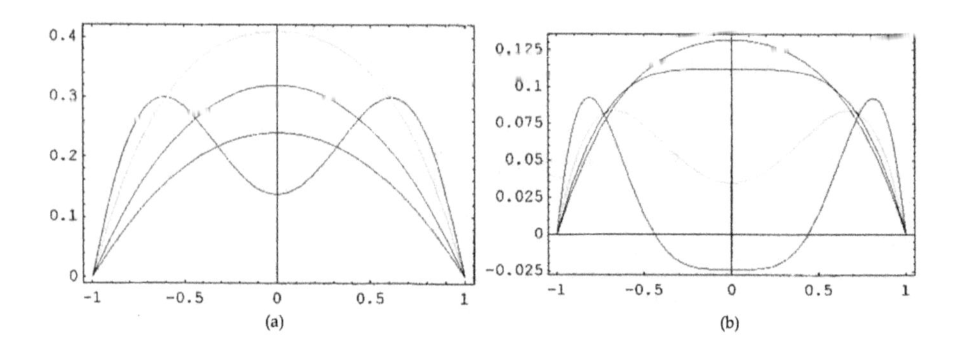

Figure 3. a) : Pulsating – APG ; n=2, H=5, β=3.7, 2.5, 1.8, 1.5 (b) : Pulsating – APG ; n=5, H=5, β=6.8, 4.1, 2.9, 2.4 [Top to Bottom for all]

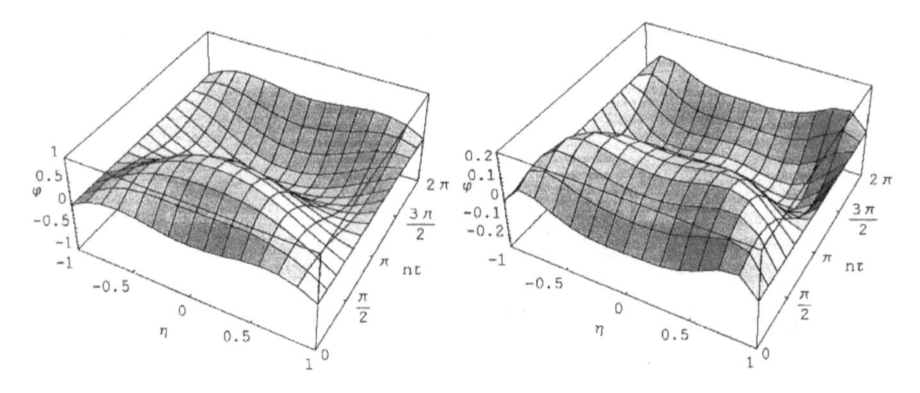

Figure 4. a): Pulsating-APG, n = 2, H = 5, at small $|\beta|$; β=3.7 (b): Pulsating-APG, n = 3, H = 5, at large $|\beta|$; β=6.8

5. Constant pressure gradient

Here we consider the flow to be initially at rest and then set in motion by a constant ABG "-K". Hence, $\Psi(\tau)$; Eq.(14), subject to BCs. (15) reduces to

$$\frac{L}{\Delta P}\frac{\partial P}{\partial z} = -K. \tag{45}$$

Therefore, we need to solve the equation

$$H\frac{\partial^2\Phi}{\partial\tau^2}+\frac{\partial\Phi}{\partial\tau}-[1+\lambda H\frac{\partial}{\partial\tau}][\frac{1}{\eta}\frac{\partial\Phi}{\partial\eta}+\frac{\partial^2\Phi}{\partial\eta^2}]=K, \quad (46)$$

subject to the boundary and initial conditions

$\phi(1,\tau)=0$, for $\tau\geq 0$,

$$\phi(\eta,0)=0,\text{for } 0\leq\eta\leq 1 \quad (47)$$

Equation (46) can be transformed to a homogenous equation by the assumption

$$\Phi(\eta,\tau)=\frac{K}{4}(1-\eta^2)-\psi(\eta,\tau), \quad (48)$$

where $\Psi(\eta,\tau)$ represents the deviation from the steady state solution. Hence,

$$[\frac{\partial}{\partial\tau}(1+H\frac{\partial}{\partial\tau})-(1+\lambda\, H\frac{\partial}{\partial\tau})(\frac{\partial^2}{\partial\eta^2}+\frac{1}{\eta}\frac{\partial}{\partial\eta})]\psi=0, \quad (49)$$

subject to the boundary and initial conditions

$$\Psi(1,\tau)=0\,\text{for}\,\tau\geq 0. \quad (50)$$

$$\psi(\eta,0)=\frac{K}{4}(1-\eta^2)\ \text{ for }\ 0\leq\eta\leq 1. \quad (51)$$

Assuming that $\psi(\eta,\tau)=F(\eta)\cdot G(\tau)$, Eq.(49) separates to

$$HG''+(1+\lambda H\alpha^2)G'+\alpha^2 G=0, \quad (52)$$

$$F''+\eta^{-1}(1+\lambda H\alpha^2)F'+\alpha^2 F=0. \quad (53)$$

Equation (52) has the solution,

$$G(\tau) = Ae^{\gamma_1 \tau} + Be^{\gamma_2 \tau} \tag{54}$$

where γ_1 and γ_2 are the roots of the Eq.(52). On the other hand, Eq. (53) has the solution

$$F(\eta) = J_0(\alpha_m \eta). \tag{55}$$

Therefore,

$$\gamma_{1,2} = \frac{-(1+\lambda H\alpha_m^2) \pm \sqrt{(1+\lambda H\alpha_m^2)^2 - 4\alpha_m^2 H}}{2H}. \tag{56}$$

The BCs. (50,51) implies that the constant α_m takes all zeros of the Bessel-function J_0 (α_1, α_2, ...). Hence,

$$\Psi(\eta,\tau) = \sum_{m=1}^{\infty} J_0(\alpha_m \eta) G(\tau), \tag{57}$$

$$\therefore \Psi(\eta,\tau) = \sum_{m=1}^{\infty} J_0(\alpha_m \eta)(A_m e^{\gamma_{1m}\tau} + B_m e^{\gamma_{2m}\tau}). \tag{58}$$

The initial condition (50) and BCs. (51) will not be sufficient to evaluate the constants A_m and B_m. Hence, it is required to employ another condition. We assume that $G(\tau)$ is smooth about the value $\tau = 0$ and can be expanded in a power series about $\tau = 0$. Assuming $G(\tau)$ to be linear function of τ in the domain about $\tau = 0$, then $G'' = 0$ in Eq. (52). Hence

$$(1+\lambda H\alpha_m^2) G_m^{`}(0) + \alpha_m^2 G_m(0) = 0, \tag{59}$$

$$G_m(\tau) = A_m e^{\gamma_{1m}\tau} + B_m e^{\gamma_{2m}\tau}, \tag{60}$$

From which we obtain

$$A_m[(1+\lambda H\alpha_m^2)\gamma_{1m} + \alpha_m^2] + B_m[(1+\lambda H\alpha_m^2)\gamma_{2m} + \alpha_m^2] = 0. \tag{61}$$

To determine the constants A_m and B_m we firstly satisfy the remaining condition (51). Owing to Eq. (58) and the initial condition, Eq. (51), we notice that,

$$\Psi(\eta,0)=\sum_{m=1}^{\infty}(A_m+B_m)J_0(\alpha_m\eta)=\frac{K}{4}(1-\eta^2). \tag{62}$$

Via the Fourier–Bessel series, Eq. (62) leads to,

$$A_m+B_m=\frac{K}{2J_1^2(\alpha_m)}\int_0^1\eta(1-\eta^2)J_0(\alpha_m\eta)d\eta. \tag{63}$$

Performing this integration we get

$$A_m+B_m=\frac{2K}{\alpha_m^3J_1(\alpha_m)}-\frac{K}{\alpha_m^2}\frac{J_0(\alpha_m)}{J_1^2(\alpha_m)}. \tag{64}$$

From Eqs. (61) and (64) we obtain :

$$A_m=\frac{[(1+\lambda H\alpha_m^2)\gamma_{2m}+\alpha_m^2]}{(1+\lambda H\alpha_m^2)(\gamma_{2m}-\gamma_{1m})}\left[\frac{2K}{\alpha_m^3J_1(\alpha_m)}-\frac{K}{\alpha_m^2}\frac{J_0(\alpha_m)}{J_1^2(\alpha_m)}\right], \tag{65}$$

$$B_m=\frac{[(1+\lambda H\alpha_m^2)\gamma_{1m}+\alpha_m^2]}{(1+\lambda H\alpha_m^2)(\gamma_{1m}-\gamma_{2m})}\left[\frac{2K}{\alpha_m^3J_1(\alpha_m)}-\frac{K}{\alpha_m^2}\frac{J_0(\alpha_m)}{J_1^2(\alpha_m)}\right]. \tag{66}$$

Finally, the velocity field has the series representation

$$\phi(\eta,\tau)=\frac{K}{4}(1-\eta^2)-\sum_{m=1}^{\infty}\frac{J_0(\alpha_m\eta)}{(1+\lambda H\alpha_m^2)(\gamma_{2m}-\gamma_{1m})}\{[(1+\lambda H\alpha_m^2)\gamma_{2m}+\alpha_m^2]e^{\gamma_{1m}\tau}$$

$$-[(1+\lambda H\alpha_m^2)\gamma_{1m}+\alpha_m^2]e^{\gamma_{2m}\tau}\}[\frac{2K}{\alpha_m^3J_1(\alpha_m)}-\frac{K}{\alpha_m^2}\frac{J_0(\alpha_m)}{J_1^2(\alpha_m)}]. \tag{67}$$

The constant-APG velocity field $\varphi(\eta,\tau)$ as a function of η shown in Fig. (5).

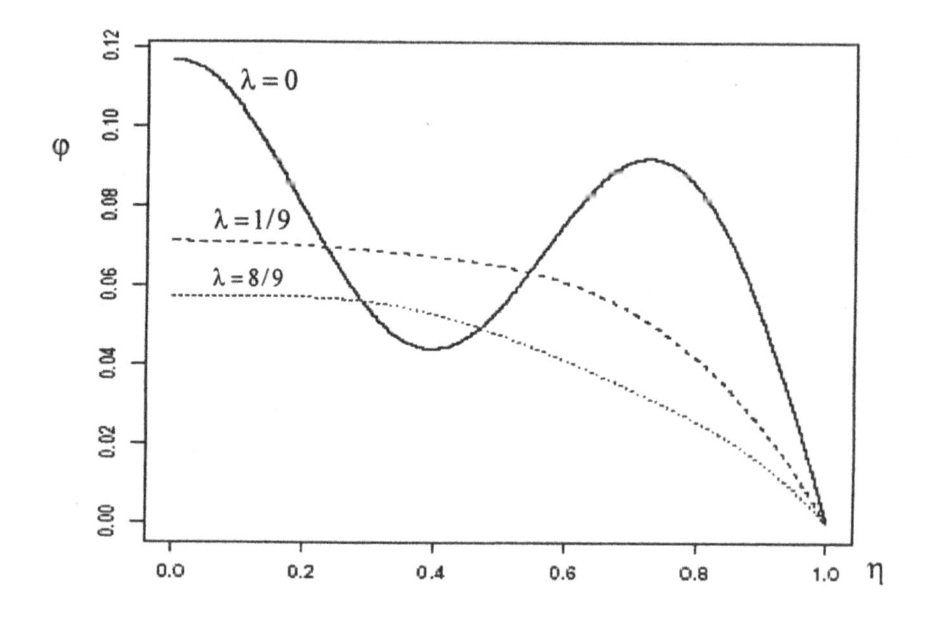

Figure 5. The velocity distribution for constant – APG taking H=0.2, τ=0.1 where the summation is taken for α_1=2.4, α_2=5.8, α_3=8.4

6. Results and discussion

The behavior of $|\beta|$ as a function of λ where H is taken as a parameter is shown in Fig. (1). The behavior of β is inversely proportional to λ while it is fast-decreasing for higher H-values. For any β-value, the Oldroyd-B fluid exhibits the same form as the UCM-fluid. A close inspection of $\beta^2 = \alpha^2(1+\alpha^2 H)/(1+\lambda\alpha^2 H)$ shows that UCM-fluid is obtained by $\lim_{\lambda\to 0}\beta^2 = \beta^2$ while $\lim_{\lambda\to 0}\beta^2 = \alpha^2$ leads to the case of Newtonian fluid. For small values of $|\beta|$ as well as $|\beta\eta|$ and by using the asymptotic expansion of $I_0(x)$, it can be shown that the velocity profiles approaches the parabolic distribution.

For decay-APGs, Figs. (2a) and (2b) show that the velocity profiles of Oldroyd-B and UCM fluids are parabolic for small values of $|\beta\eta|$ while for large $|\beta\eta|$ they are completely different from this situation. The solutions depend on η only in the neighboring of the wall. Therefore, such fluids exhibit boundary layer effects [17]].

For pulsating-APG, the velocity distribution is represented in Figs. (3a) and (3b). The smallest value of β in both curves is almost parabolic as shown by Eq. (36) while the largest value exhibits boundary effect as reviled by Eq.(43). To emphasize the oscillating nature of the solution a three-dimensional diagrams (4a) and (4b) for the smallest and largest values of $|\beta|$ are respectively sketched.

Grigioni, et al [1], studided the behavior of blood as a viscoelastic fluid using the Oldroyd-B model. The results obtained for the velocity distribution stands in agreement with the obtained results in the present work.

Author details

A. Abu-El Hassan and E. M. El-Maghawry*

*Address all correspondence to: abd_galil@hotmail.com

Physics Department, Faculty of Science, Benha University, Egypt

References

[1] Grigioni, M., Daniele, C., and D' Avenio, G. The Role of Wall Shear Stress in Unsteady Vascular Dynamics, Vol. 7, No.3 Sep. (2002).

[2] Thurston, G. B., The viscoelasticity of human blood, Biophysical Journal, 12, 1205-1217(1972).

[3] Perspectives in fluid dynamics,A collective introduction to current research, Edited by: Batchelor, G.K., Moffatt, H.K. and Worster, M.G., Cambridge Univ. Press,(2000).

[4] Tucker, C. L., Fundamentals of computer modeling for polymer processing, III, HANSER Publishers, Munich (1989).

[5] Bird, R.B., Curtiss.C.F., Armstrong R.C. and Hassager, O., Dynamics of polymeric liquids, Vol.2, Wiley, New York (1987).

[6] Rahaman, K. D. and Ramkissoon, H., Unsteady axial viscoelastic pipe flows, J.Non-Newtonian Fluid Mech., 57 (1995) 27.

[7] Rajagopal, K.R., Int. J. Non-linear Mech. 17 (1982) 369.

[8] Atalik, K. and R. Keunings,R., J. Non-Newtonian Fluid mech. 102, (2002) 299.

[9] Yesilata, B, Fluid Dyn. Res. 31, 41 (2002).

[10] Pontrelli, G., Pulsatile blood flow in a pipe. Computers & Fluids, 27 (1998) 367.

[11] Pontrelli, G, Blood flow through a circular pipe with an impulsive pressure gradient. Math Models Methods in Appi Sci, 10 (2000) 187

[12] Bames, H.A., Townsend, P. and Walters, K., Rheol. Acta.10 (1971) 517.

[13] Bames, H.A., Townsend, P. and Walters, K., 244, Nature (1969) 585.

[14] Davies, J.M., Bhumiratana, S. and Bird, R.B. J. Non-Newtonian Fluid Mech, 3 (1977/1978) 237.

[15] Phan-Thien, N. and DudekJ, J.Non-Newtonian Fluid Mech., 11 (1982) 147.

[16] Hayat, T, Asghar, S. and Siddiqui, A.M., Some unsteady unidirectional flows of a Non -Newtonian fluid, Int. J. Eng. Science, 38 (2000) 337.

[17] Schlichting, H., Boundary- layer theory, McGraw-Hill, New York, 1968

Chapter 6

Measurement and Prediction of Fluid Viscosities at High Shear Rates

Jeshwanth K. Rameshwaram and Tien T. Dao

Additional information is available at the end of the chapter

1. Introduction

Polymeric materials in general are viscoelastic in nature because they exhibit strong dependence of deformation and flow on time and temperature. Molecular structures in polymeric materials undergo rearrangements when a load is applied in order to minimize localized stresses imposed by the applied load. Performance of materials in real time applications can only be evaluated by testing the materials under conditions that the material would encounter in real time applications. This poses serious problems in evaluating materials that would need to be in service over very long periods of time or if the material undergoes deformations (rates) over a wide range. For example, it would be impossible to evaluate the flow behavior of materials in solar panels over the span of time they are designed for (20 – 30 years). Similarly, aircraft/spacecraft materials that are rated for service up to thousands of hours cannot be evaluated. Measurements taken using an instrument at a set temperature usually cover a range of three to four orders of magnitude of time or frequency [1]. This range is usually not sufficient to evaluate material viscoelastic behavior in the complete range of frequencies required, from the low frequency terminal zone to high frequency glassy region.

Viscosity measurements of materials have become a very essential part of a wide variety of industries including petroleum, food, plastics, paint and composite industries, especially in the last decade. With the advent of advanced lightweight structures like fiber reinforced plastics, that possess high specific strength and specific modulus, the development of testing techniques to evaluate the lifetimes of such materials in operating environments has become of high importance. The life times of some of these materials are a few years which makes it impossible to conduct real time experiments for that span of time. The TTS technique works very well in such situations to predict the behavior of the material over the life time of the material in specific applications.

Time-temperature superposition (TTS) has been used to solve the kind of problems mentioned above over the last few decades. This technique is well grounded in theory and applies to a wide variety of hydrocarbon materials that are thermorheologically simple. Thermorheologically simple materials are those where all relevant relaxation and retardation mechanisms as well as stress magnitudes at all times and frequencies have the same temperature dependence.

The theory and mechanisms that form the basis for the TTS technique have been well documented in literature. However, a brief overview of the underlying theory of the TTS technique is presented to refresh the readers' memory.

The first step towards using the TTS technique is to generate data at several temperatures close to the temperature at which the flow properties/deformations of the material in question are to be evaluated (reference temperature). The flow curves at the temperatures obtained can then be superposed on to a master curve showing the material behavior at reference temperature. The TTS technique is based on the facts that the molecular rearrangements that occur due to applied stresses take place at accelerated rates at elevated temperatures and that there is an analogous relationship between time and temperature [2, 3]. These phenomena lend to the ability to conduct measurements at elevated temperatures and then superpose the data to lower temperatures. Superposing the data to lower temperatures enables the analyst to predict the material behavior over a large time scales (small deformations over long periods of time). Similarly, superposing lower temperature data to a higher temperature enables one to predict the material behavior over smaller time scales (high shear rates).

The extent of shifting along the x and y axes in the TTS technique, in order to superpose experimental data on to that at the reference temperature, is represented by the horizontal shift factor a_T and vertical shift factor b_T. Variables that have units of time or reciprocal time get subjected to a horizontal shift and variables that have units of stress or reciprocal stress are subject to a vertical shift [1]. The vertical shift factor (b_T) can be calculated using the equation 1 [4, 5]:

$$b_T = T_0 \varrho_0 / T \varrho \tag{1}$$

The horizontal shift factor is given by the Arrhenius relationship:

$$a_T\left(T\right) = exp\left[\frac{E_a}{R}\left(\frac{1}{T} - \frac{1}{T_0}\right)\right] \tag{2}$$

Where E_a is the activation energy, R is the gas constant, T is the measurement temperature and T_0 is the reference temperature. This relationship is valid as long as the measurement temperature is well above the T_g of the material. For measurement temperatures closer to T_g, the following relationship holds well:

$$\log a_T = \frac{-c_1(T - T_0)}{[c_2 + (T - T_0)]} \tag{3}$$

Where c_1 and c_2 are empirical constants.

The application of TTS to predict the creep behavior of materials over long period of time has been in practice for a few decades. However, predicting material flow behavior and viscosity values of materials at extremely high deformation rates is a relatively unchartered territory. This study demonstrates the use of capillary rheology to measure viscosity profiles of motor oils up to high shear rates of 2,000,000 s^{-1}. Viscosity profiles of oils at six different temperatures up to shear rates of ~ 2,000,000 $s^{-1}$are presented. The data sets at lower temperatures (100, 110, 120, 130 and 140 °C) are superposed on to the data set at 150 °C using the TTS technique (IRIS software) in order to predict the viscosity behavior of the oil at extremely high shear rates (up to ~ 15 million s^{-1}) at 150 °C. This study demonstrates that the TTS technique can be used to predict such data with reasonable accuracy.

2. Experimental

Steady shear rheological measurements of two [2] oil samples: Newtonian control oil sample and a commercially available multiweight motor oil, using a RH 2000 Dual Bore, Bench Top Capillary Rheometer with a die of dimensions 0.156 x 65 mm are presented in this study. A 69 MPa pressure transducer is used for all experiments in this study. The shear rates used in this study range from ~ 50,000 – 2,000,000 s^{-1}. Viscosity profiles of both oils at 100, 110, 120, 130, 140 and 150 °C are presented.

The viscosity profiles of the control oil show Newtonian behavior at all temperatures and shear rates measured. The commercial oil however, exhibits slight shear thinning at lower temperatures and Newtonian behavior at higher temperatures. Also, the viscosities of both samples decrease with increase in temperature. The dies used in this study are fitted in an in-house designed die holder with a heating sleeve heated to test temperatures in order to sustain uniform operating temperatures throughout the length of the dies.

Time temperature superposition (TTS) is used to estimate viscosity values of the two [2] oils at extremely high shear rates using the IRIS software. TTS shifting gives viscosity data at higher shear rates with reasonable accuracy while eliminating the upturn in viscosity at high shear rates due to turbulence. Viscosity profiles of respective oils at 100, 110, 120, 130 and 140 °C are shifted to superpose on to the viscosity profile at 150 ºC to generate the master curve, which predicts the viscosity profile up to extremely high shear rates (~ 15 million s^{-1}).

3. Results and discussion

The reliability of any data always depends on the calibration and accurate measurement capabilities of the instrument. Therefore, proper calibration and verification of the performance of the instrument is necessary in order to generate accurate and reliable data. Figure 1 presents the viscosity profile of a NIST traceable Newtonian viscosity standard oil with nominal viscosity of 10 cP at 25 ºC. The data verifies the accuracy of the measurements and demonstrates that the instrument is properly calibrated.

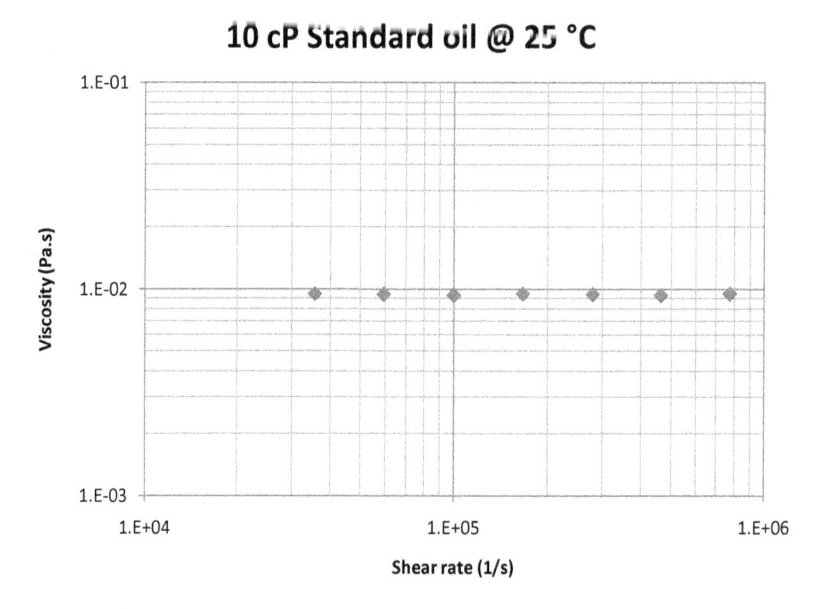

Figure 1. Steady state shear viscosity of 10 cP nominal viscosity standard oil at 25°C.

Figure 2 presents the viscosity profiles of a Newtonian oil at 100, 110, 120, 130, 140 and 150 °C. Data indicates that the oil is Newtonian in the range of shear rates and temperatures measured and is sensitive to temperature, i.e., viscosity decreases with increase in temperature. The high end shear rates measured experimentally are unique to this study. Viscosity measurements up to the high shear rates represented in this study are achievable by using a die with high L/D (length/diameter) ratio (> 400) using an in-house design. The die fits into an in-house designed die holder, which can be modified to fit any commercially available capillary rheometer. Commercially available capillary rheometers are generally not equipped to measure viscosities of low viscosity fluids at high shear rates. The temperature range used in this study is chosen due to the fact that flow behavior of the oils at temperatures below 100 ºC indicates strong shear thinning at high shear rates. A shear thinning viscosity profile cannot be superposed on to a Newtonian viscosity profile to generate the master curve using the TTS technique because the relaxation and retardation mechanisms at play in shear thinning behavior are different from the mechanisms at play in Newtonian behavior. As mentioned above, only data that involves similar molecular rearrangement mechanisms can be used to conduct the time-temperature superposition.

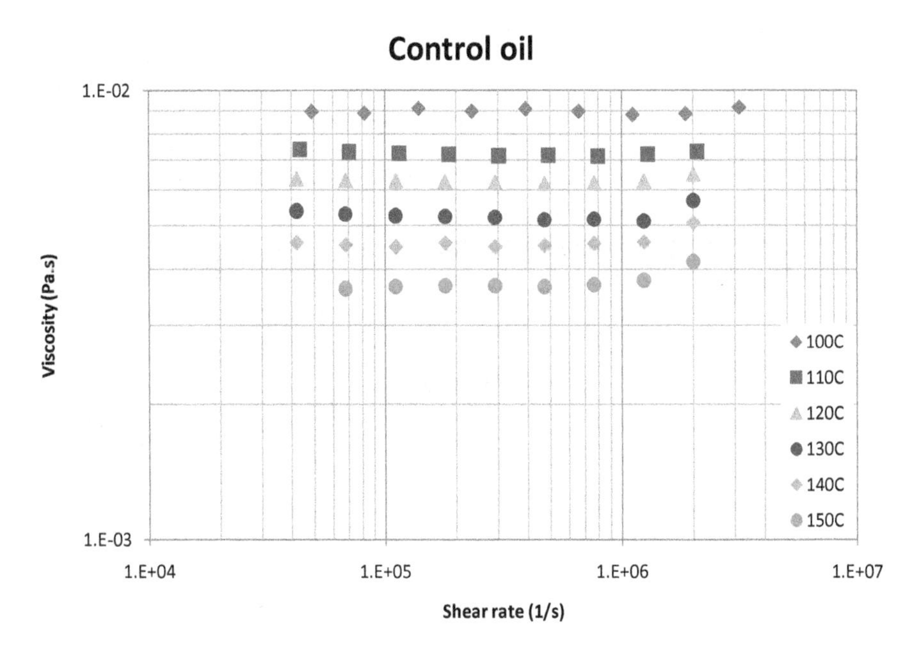

Figure 2. Steady state shear viscosity profiles of the control oil at 100, 110, 120, 130, 140 and 150 °C.

The commercial multiweight oil viscosity profiles in the temperature range of 100 – 150 ºC at 10 ºC increments are presented in Figure 3. Data indicates that this oil is slightly shear thinning up to 130 ºC, but shows Newtonian behavior above this temperature. As expected, the oil viscosity decreases with increase in temperature. It is to be noted that in Figures 2 and 3, the viscosity profiles at 120 ºC and above show an upturn at shear rates above 1,000,000 s^{-1}. This upturn is due to the onset of secondary flows (turbulence). Turbulent flow manifests itself as an upturn in viscosity due to additional resistance to flow. The use of the high L/D ratio dies pushes the onset of turbulence to shear rates of ~ 1,000,000 s^{-1}, which otherwise would set in at much lower shear rates leading to misleading viscosity values. This in turn would make it impossible to estimate viscosity values at high shear rates using TTS.

One of the applications of this study is in estimating the viscosity of motor oil in engine parts, where the oil undergoes extremely high shear rates. As there is no way of achieving true viscosity values experimentally while mimicking the shear rates present in the engine parts, the use of TTS to estimate the viscosity values at such high shear rates (~ 1,000,000 – 15,000,000 s^{-1}) is very useful. Also, the detrimental effects of turbulent flow in viscosity measurements at high shear rates can be eliminated by estimating the viscosity values using TTS.

Figure 4 presents the master curves obtained by superposing the viscosity profiles of the control sample oil and the commercial multiweight non-Newtonian oils (from Figures 2 & 3, respectively) at 100 – 140 ºC on to that at 150 ºC. As expected, estimated viscosity values using TTS at shear rates higher than those measured experimentally (> 1,000,000 s^{-1}) show that the Newtonian oil is constant up to ~ 15,000,000 s^{-1} shear rate. However, the non-Newtonian oil exhibits slight shear thinning behavior, as would be expected from the experimental data. Figure 4 also compares the experimental data sets at 150 ºC with respective master curves generated using TTS. It is demonstrated that the upturn in viscosity due to secondary flows is eliminated as well by using the TTS. The *y-axis* is plotted on a linear scale for clarity.

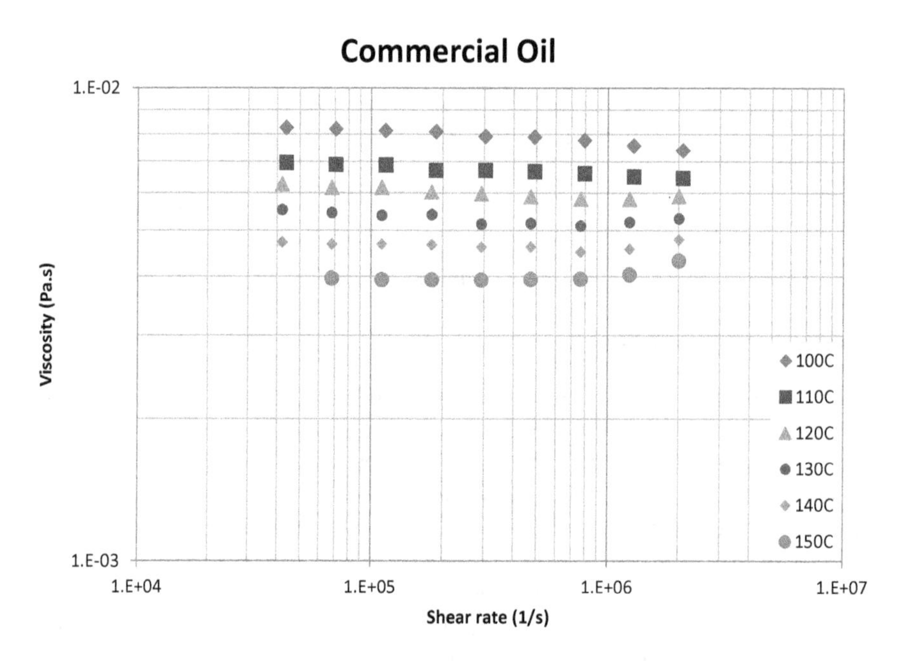

Figure 3. Steady state shear viscosity profiles of the commercial multiweight oil sample at 100, 110, 120, 130, 140 and 150 °C.

The transition from laminar to turbulent flow in capillary rheometry plays an important role in determining whether the viscosity values obtained experimentally are real. This transition can be determined by calculating the Reynolds numbers. Figure 5 presents the Reynolds numbers of the control oil as a function of pressure drop. The Reynolds numbers are calculated using the Fanning friction factor given in equation (4) [6]:

$$f = \frac{\Delta P}{L} \times \frac{D}{2\varrho v^2} \tag{4}$$

where ΔP is the pressure drop, D is the die diameter, L is the die length, ϱ is the fluid density and v is the fluid velocity in the die. In order to check if the flow in the studied shear rate range is laminar, the Reynolds numbers are also calculated using the Hagen-Poiseuille equation given by equation (5):

$$f = \frac{16}{R_e} \left(\text{Laminar flow}\right) \tag{5}$$

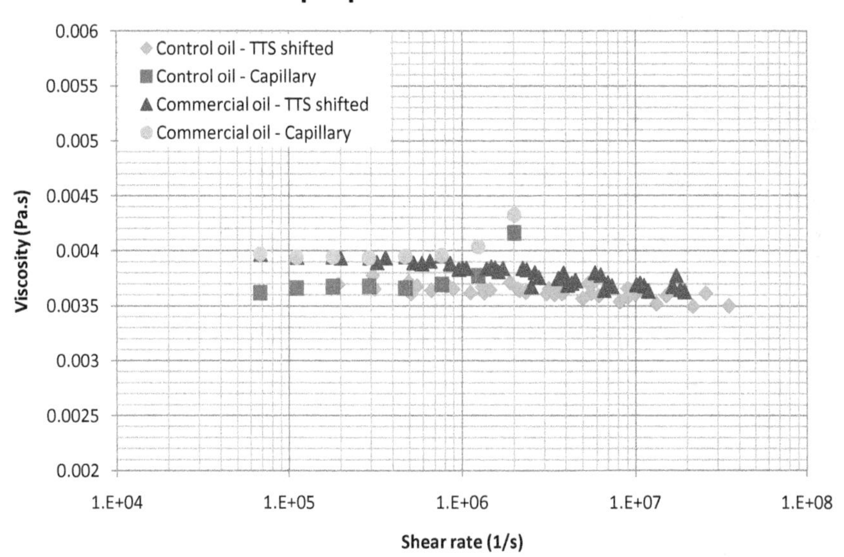

Figure 4. Estimated viscosity data of the control and commercial oils over a very large range of shear rates at 150 °C using TTS compared to original viscosity profile using capillary rheology.

For flow in the transition and turbulent regions, the Blasius formula given in equation (6) is employed:

$$f = \frac{0.079}{R_e^{\frac{1}{4}}} \left(\text{Turbulent flow}\right) \tag{6}$$

Reynolds numbers for all samples are calculated using equations (5) and (6) to check if the flow is in the laminar region or the turbulent region. Reynolds number calculations suggest there is a transition in the flow pattern from laminar to turbulent flow in the Reynolds number range of 850 – 1000. The pressure drop data is corrected for kinetic energy contributions as well. From experience, the transition from laminar to turbulent flow commences when the graphs of laminar and turbulent flow start to converge, in this case around a Reynolds number of ~ 850. The onset of complete turbulent flow is assumed to set in approximately at the point where the laminar flow graph intersects the turbulent flow graph. The *x-axis* in the figure is plotted on a log scale for clarity.

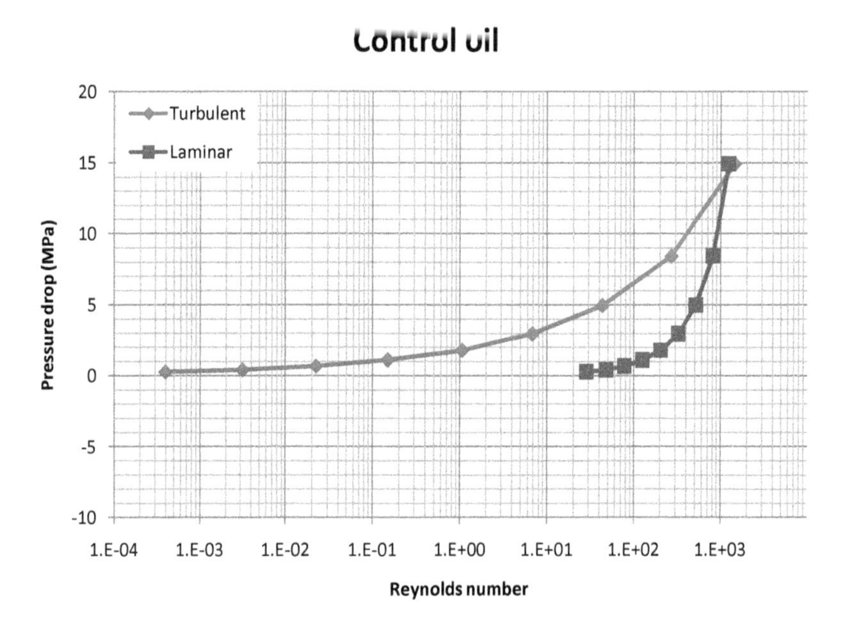

Figure 5. Reynolds numbers of the control oil as a function of pressure drop using equations (b) and (c).

Figure 6 shows the Reynolds number at the point of viscosity upturn for the control oil at high shear rates. This upturn in viscosity is attributed to the transition from laminar flow to turbulence in capillary flow for this configuration. For the purpose of this study, an increase in viscosity by ~ 10% is where the transition from laminar to turbulent flow is assumed to commence. The Reynolds number of ~ 820 presented in Figure 6 is calculated using equation (5), and the Reynolds number 1450 is calculated using equation (6).

In order to check for viscous heating effects in the capillary at high shear rates, the Nahme numbers (*Na*) for all the data points presented in this report are calculated. The *Na* number needs to be less than 1 for die viscous heating to be neglected. The *Na* number can be calculated by the following equation (7): where β is the temperature sensitivity of viscosity, η is the viscosity of the fluid, γ is the shear rate, R is the radius of the die, and k is the thermal conductivity of the fluid.

$$Na = \frac{\beta \eta \gamma R^2}{4k} \tag{7}$$

All Na values except at shear rates above 1,000,000 s^{-1} were lower than 1 for both oils, thereby indicating that viscous heating effects can be neglected up to shear rates of 1,000,000 s^{-1}. Data above 1,000,000 s^{-1}shear rate at any temperature is not used for TTS shifting in this study. It should be noted that the Na approaches 1 at 1,000,000 s^{-1} at 100 ºC. Samples with higher viscosity and/or at lower temperatures will have Na higher than 1 below 1,000,000 s^{-1} shear rates, in which case, viscous heating effects would need to be taken into account while calculating the real viscosities of such samples.

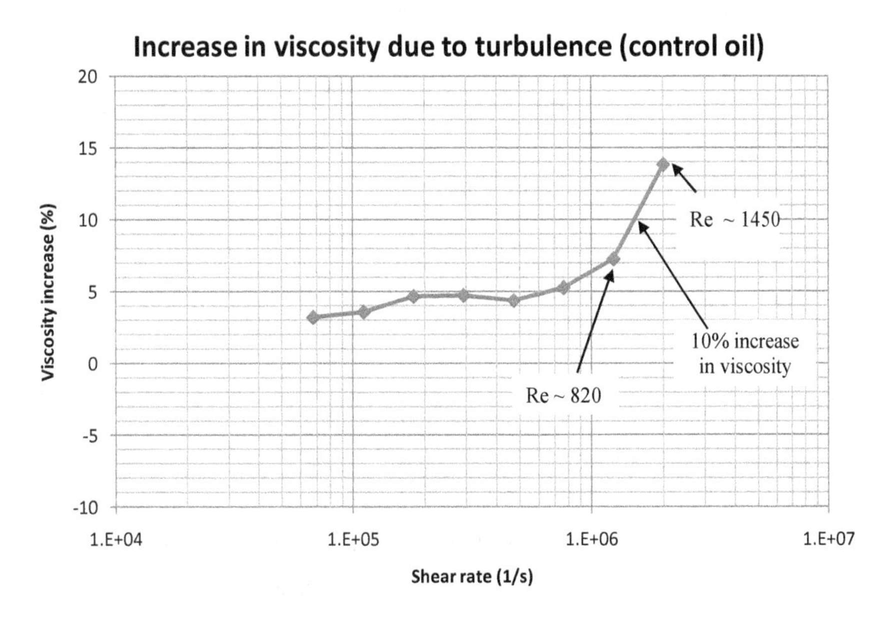

Figure 6. Increase in viscosity (percent) of the control oil due to onset of turbulence at 150 °C.

4. Conclusions

This study emphasizes the importance and explains the applications of high shear rate viscosity measurements. A novel technique to measure low viscosity fluids at high shear rates (~ 50,000 - 2,000,000 s^{-1}) is developed using capillary rheology. The difficulty to achieve reliable viscosity measurements at shear rates above 1,000,000 s^{-1} is addressed and a method (time-temperature superposition) to estimate the viscosity behavior of low viscosity fluids with reasonable accuracy up to extremely high shear rates (~ 15,000,000 s^{-1}) is established. It is demonstrated that the time-temperature superposition (TTS) technique also alleviates the problem of errant viscosity numbers at high shear rates (> 1,000,000 s^{-1}) due to onset of turbulent flow. This study is expected to have a prolific impact on low viscosity - high shear rate applications such as oil flowing through engine parts, paint spray guns, petroleum pipelines, among other applications in various industries.

Acknowledgements

We sincerely appreciate all the cooperation extended to us by the author of the IRIS software, Dr. Henning Winter, regarding applying the time-temperature superposition to extend the shear rate range in this study using the IRIS software.

Author details

Jeshwanth K. Rameshwaram* and Tien T. Dao

*Address all correspondence to: jkr@atsrheosystems.com

ATS RheoSystems, Bordentown, NJ, USA

References

[1] Dealy, J and Plazek, D. Time-temperature Superposition – A users guide. Rheology Bulletin, 78(2), 16 – 31 (2009).

[2] A.V. Tobolsky, Properties and Structures of Polymers, Wiley, New York, (1960).

[3] J.D. Ferry, Viscoelastic Properties of Polymers, 3rd ed., John Wiley & Sons, NY. (1980).

[4] P.E. Rouse, "A theory of linear viscoelastic properties of dilute solutions of coiling polymers," J. of Chem. Phys, 21, 1272 – 1280 (1953).

[5] F. Bueche, " Viscosity self diffusion and allied effect in solid polymers," J. Chem Phys, 20, 1959 – 1964 (1952).

[6] Bird et al., Transport Phenomena, 2ed, John Wiley and Sons, NY, pp 179 to 183 (2001)

Permissions

The contributors of this book come from diverse backgrounds, making this book a truly international effort. This book will bring forth new frontiers with its revolutionizing research information and detailed analysis of the nascent developments around the world.

We would like to thank Rajkumar Durairaj, for lending his expertise to make the book truly unique. He has played a crucial role in the development of this book. Without his invaluable contribution this book wouldn't have been possible. He has made vital efforts to compile up to date information on the varied aspects of this subject to make this book a valuable addition to the collection of many professionals and students.

This book was conceptualized with the vision of imparting up-to-date information and advanced data in this field. To ensure the same, a matchless editorial board was set up. Every individual on the board went through rigorous rounds of assessment to prove their worth. After which they invested a large part of their time researching and compiling the most relevant data for our readers. Conferences and sessions were held from time to time between the editorial board and the contributing authors to present the data in the most comprehensible form. The editorial team has worked tirelessly to provide valuable and valid information to help people across the globe.

Every chapter published in this book has been scrutinized by our experts. Their significance has been extensively debated. The topics covered herein carry significant findings which will fuel the growth of the discipline. They may even be implemented as practical applications or may be referred to as a beginning point for another development. Chapters in this book were first published by InTech; hereby published with permission under the Creative Commons Attribution License or equivalent.

The editorial board has been involved in producing this book since its inception. They have spent rigorous hours researching and exploring the diverse topics which have resulted in the successful publishing of this book. They have passed on their knowledge of decades through this book. To expedite this challenging task, the publisher supported the team at every step. A small team of assistant editors was also appointed to further simplify the editing procedure and attain best results for the readers.

Our editorial team has been hand-picked from every corner of the world. Their multi-ethnicity adds dynamic inputs to the discussions which result in innovative outcomes. These outcomes are then further discussed with the researchers and contributors who give their valuable feedback and opinion regarding the same. The feedback is then collaborated with the researches and they are edited in a comprehensive manner to aid the understanding of the subject.

Apart from the editorial board, the designing team has also invested a significant amount of their time in understanding the subject and creating the most relevant covers. They scrutinized every image to scout for the most suitable representation of the subject and create an appropriate cover for the book.

The publishing team has been involved in this book since its early stages. They were actively engaged in every process, be it collecting the data, connecting with the contributors or procuring relevant information. The team has been an ardent support to the editorial, designing and production team. Their endless efforts to recruit the best for this project, has resulted in the accomplishment of this book. They are a veteran in the field of academics and their pool of knowledge is as vast as their experience in printing. Their expertise and guidance has proved useful at every step. Their uncompromising quality standards have made this book an exceptional effort. Their encouragement from time to time has been an inspiration for everyone.

The publisher and the editorial board hope that this book will prove to be a valuable piece of knowledge for researchers, students, practitioners and scholars across the globe.

List of Contributors

Bradley W. Mansel and Philipus J. Patty
Institute of Fundamental Sciences, Massey University, Palmerston North, New Zealand

Yacine Hemar
MacDiarmid Institute for Advanced Materials and Nanotechnology, New Zealand
School of Chemical Sciences, University of Auckland, New Zealand

Martin A.K. Williams
Institute of Fundamental Sciences, Massey University, Palmerston North, New Zealand
MacDiarmid Institute for Advanced Materials and Nanotechnology, New Zealand
Riddet Institute, Palmerston North, New Zealand

Stephen Keen
Institute of Fundamental Sciences, Massey University, Palmerston North, New Zealand
MacDiarmid Institute for Advanced Materials and Nanotechnology, New Zealand

Trofimov Alexander and Sevostyanova Evgeniya
International Scientific Research Institute of Cosmic Anthropoecology, Scientific Center of Clinical and Experimental Medicine of SB RAMS, Novosibirsk, Russia

R. Durairaj, Lam Wai Man, Kau Chee Leong, Liew Jian Ping and Lim Seow Pheng
Department of Mechanical and Material Engineering, Faculty of Engineering and Science (FES), Universiti Tunku Abdul Rahman (UTAR), Jalan Genting Kelang, Setapak, Kuala Lumpur, Malaysia

N. N. Ekere
School of Technology, University of Wolverhampton, Technology Centre (MI Building), City Campus - South, Wulfruna St, Wolverhampton, United Kingdom

R. Talero and C. Pedrajas
Eduardo Torroja Institute for Construction Sciences - CSIC; Calle Serrano Galvache; Madrid, Spain

V. Rahhal
Departamento de Ingeniería Civil Facultad de Ingeniería UNCPBA, Av. del Valle, Argentina, Olavarría, Argentina

A. Abu-El Hassan and E. M. El-Maghawry
Physics Department, Faculty of Science, Benha University, Egypt

Jeshwanth K. Rameshwaram and Tien T. Dao
ATS RheoSystems, Bordentown, NJ, USA